PROBLEM
SOLVING
in
MATHEMATICS

Cover design by Bev and Charles Dana

This work was developed under an ESEA Title IVC grant from the Oregon Department of Education, Office of Policy and Program Development. The content, however, does not necessarily reflect the position or policy of the Oregon Department of Education and no official endorsement of these materials should be inferred.

Distribution for this work was arranged by LINC Resources, Inc.

ISBN 0-86651-183-0

Order Number DS01409

12 13 14 15 16-MA-99 98

DALE
SEYMOUR
PUBLICATIONS
P.O. BOX 10888
PALO ALTO, CA 94303

PROBLEM SOLVING IN MATHEMATICS

PROJECT STAFF

DIRECTOR: OSCAR SCHAAF, UNIVERSITY OF OREGON
ASSOCIATE DIRECTOR: RICHARD BRANNAN, LANE EDUCATION SERVICE DISTRICT

WRITERS: RICHARD BRANNAN
                         MARYANN DEBRICK
                         JUDITH JOHNSON
                         GLENDA KIMERLING
                         SCOTT McFADDEN
                         JILL McKENNEY
                         OSCAR SCHAAF
                         MARY ANN TODD

PRODUCTION: MEREDITH SCHAAF
                         BARBARA STOEFFLER

EVALUATION: HENRY DIZNEY
                         ARTHUR MITTMAN
                         JAMES ELLIOTT
                         LESLIE MAYES
                         ALISTAIR PEACOCK

PROJECT GRADUATE   FRANK DEBRICK
   STUDENTS: MAX GILLETT
                         KEN JENSEN
                         PATTY KINCAID
                         CARTER McCONNELL
                         TOM STONE

ACKNOWLEDGEMENTS:

TITLE IV-C LIAISON: Ray Talbert
                              Charles Nelson

                              Monitoring Team

                              Charles Barker
                              Ron Clawson
                              Jeri Dickerson
                              Anthony Fernandez
                              Richard Olson
                              Ralph Parrish
                              Fred Rugh
                              Alton Smedstad

ADVISORY COMMITTEE:

Mary Grace Kantowski	University of Florida
John LeBlanc	Indiana University
Richard Lesh	Northwestern University
Edwin McClintock	Florida International University
Len Pikaart	Ohio University
Kenneth Vos	The College of St. Catherine

A special thanks is due to the many teachers, schools, and districts within the state of Oregon that have participated in the development and evaluation of the project materials. A list would be lengthy and certainly someone's name would inadvertently be omitted. Those persons involved have the project's heartfelt thanks for an impossible job well done.

The following projects and/or persons are thanked for their willingness to share pupil materials, evaluation materials, and other ideas.

    Don Fineran, Mathematics Consultant, Oregon Department of Education
    Frank Lester, Indiana University
    Steve Meiring, Mathematics Consultant, Ohio Department of Education
    Harold Schoen, University of Iowa
    Iowa Problem Solving Project, Earl Ockenga, Manager
    Math Lab Curriculum for Junior High, Dan Dolan, Director
    Mathematical Problem Solving Project, John LeBlanc, Director

CONTENTS

INTRODUCTION

What is PSM?

PROBLEM SOLVING IN MATHEMATICS is a program of problem-solving lessons and teaching techniques for grades 4–8 and (9) algebra. Each grade-level book contains approximately 80 lessons and a teacher's commentary with teaching suggestions and answer key for each lesson. *Problem Solving in Mathematics* is not intended to be a complete mathematics program by itself. Neither is it supplementary in the sense of being extra credit or to be done on special days. Rather, it is designed to be integrated into the regular mathematics program. Many of the problem-solving activities fit into the usual topics of whole numbers, fractions, decimals, percents, or equation solving. Each book begins with lessons that teach several problem-solving skills. Drill and practice, grade-level topics, and challenge activities using these problem-solving skills complete the book.

PROBLEM SOLVING IN MATHEMATICS is designed for use with all pupils in grades 4–8 and (9) algebra. At-grade-level pupils will be able to do the activities as they are. More advanced pupils may solve the problems and then extend their learning by using new data or creating new problems of a similar nature. Low achievers, often identified as such only because they haven't reached certain computational levels, should be able to do the work in PSM with minor modifications. The teacher may wish to work with these pupils at a slower pace using more explanations and presenting the material in smaller doses.

[Additional problems appropriate for low achievers are contained in the *Alternative Problem Solving in Mathematics* book. Many of the activities in that book are similar to those in the regular books except that the math computation and length of time needed for completion are scaled down. The activities are generally appropriate for pupils in grades 4–6.]

Why Teach Problem Solving?

Problem solving is an ability people need throughout life. Pupils have many problems with varying degrees of complexity. Problems arise as they attempt to understand concepts, see relationships, acquire skills, and get along with their peers, parents, and teachers. Adults have problems, many of which are associated with making a living, coping with the energy crisis, living in a nation with peoples from different cultural backgrounds, and preserving the environment. Since problems are so central to living, educators need to be concerned about the growth their pupils make in tackling problems.

What Is a Problem?

MACHINE HOOK-UPS

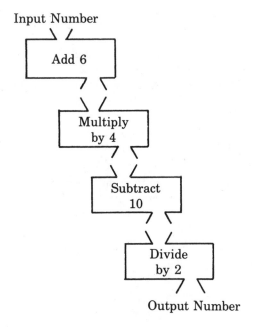

It is highly recommended that teachers intending to use *Problem Solving in Mathematics* receive training in implementing the program. The *In-Service Guide* contains much of this valuable material. In addition, in-service audio cassette tapes are available. These provide indepth guidance on using the PSM grade-level books and an overall explanation of how to implement the whole program. The tapes are available for loan upon request. Please contact Dale Seymour Publications, Box 10888, Palo Alto, CA 94303 for further information about the tapes and other possible in-service opportunities.

	Input Number	Output Number
a.	4	
b.	8	
c.	12	
d.		39
e.		47
f.		61

Suppose a 6th grader were asked to fill in the missing output blanks for *a*, *b*, and *c* in the table. Would this be a problem for him? Probably not, since all he would need to do is to follow the directions. Suppose a 2nd-year algebra student were asked to fill in the missing input blank for *d*. Would this be a problem? Probably not, since she would write the suggested equation,

$$\frac{4\,(x + 6) - 10}{2} = 39$$

and then solve it for the input. Now suppose the 6th grader were asked to fill in the input for *d*, would this be a problem for him? Probably it *would* be. He has no directions for getting the answer. However, if he has the desire, it is within his power to find the answer. What might he do? Here are some possibilities:

1. He might make *guesses*, do *checking*, and then make refinements until he gets the answer.
2. He might fill in the output numbers that correspond to the input numbers for *a*, *b*, and *c*.

	Input Number	Output Number
a.	4	
b.	8	
c.	12	
d.		39

and then observe this pattern:
For an increase of 4 for the input, the output is increased by 8.
Such an observation should lead quickly to the required input of 16.

3. He might start with the output and *work backwards* through the machine hook-up using the inverse (or opposite) operations.

For this pupil, there was no "ready-made" way for him to find the answer, but most motivated 6th-grade pupils would find a way.

A *problem*, then, is a situation in which an individual or group accepts the challenge of performing a task for which there is no immediately obvious way to determine a solution. Frequently, the problem can be approached in many ways. Occasionally, the resulting investigations are nonproductive. Sometimes they are so productive as to lead to many different solutions or suggest more problems than they solve.

What Does Problem Solving Involve?

Problem solving requires the use of many *skills*. Usually these skills need to be used in certain combinations before a problem is solved. A combination of skills used in working toward the solution of a problem can be referred to as a *strategy*. A successful strategy requires the individual or group to generate the information needed for solving the problem. A considerable amount of creativity can be involved in generating this information.

What Problem-Solving Skills Are Used in PSM?

Skills are the building blocks used in solving a problem. The pupil materials in the PSM book afford many opportunities to emphasize problem-solving skills. A listing of these skills is given below.

THE PSM CLASSIFIED LIST
OF PROBLEM-SOLVING SKILLS

A. Problem Discovery, Formulation
   1. State the problem in your own words.
   2. Clarify the problem through careful reading and by asking questions.
   3. Visualize an object from its drawing or description.
   4. Follow written and/or oral directions.

B. Seeking Information
5. Collect data needed to solve the problem.
6. Share data and results with other persons.
7. Listen to persons who have relevant knowledge and experiences to share.
8. Search printed matter for needed information.
9. Make necessary measurements for obtaining a solution.
10. Record solution possibilities or attempts.
11. Recall and list related information and knowledge.

C. Analyzing Information
12. Eliminate extraneous information.
13. Find likenesses and differences and make comparisons.
14. Classify objects or concepts.
15. Make and use a drawing or model.
16. Make and/or use a systematic list or table.
17. Make and/or use a graph.
18. Look for patterns and/or properties.
19. Use mathematical symbols to describe situations.
20. Break a problem into manageable parts.

D. Solve—Putting It Together—Synthesis
21. Make predictions, conjectures, and/or generalizations based upon data.
22. Make decisions based upon data.
23. Make necessary computations needed for the solution.
24. Determine limits and/or eliminate possibilities.
25. Make reasonable estimates.
26. Guess, check, and refine.
27. Solve an easier but related problem. Study solution process for clues.
28. Change a problem into one you can solve. (Simplify the problem.)
29. Satisfy one condition at a time.
30. Look at problem situation from different points of view.
31. Reason from what you already know. (Deduce.)
32. Work backwards.
33. Check calculated answers by making approximations.
34. Detect and correct errors.
35. Make necessary measurements for checking a solution.

36. Identify problem situation in which a solution is not possible.
37. Revise the conditions of a problem so a solution is possible.

E. Looking Back—Consolidating Gains
38. Explain how you solved a problem.
39. Make explanations based upon data.
40. Solve a problem using a different method.
41. Find another answer when more than one is possible.
42. Double check solutions by using some formal reasoning method (mathematical proof).
43. Study the solution process.
44. Find or invent other problems which can be solved by certain solution procedures.
45. Generalize a problem solution so as to include other solutions.

F. Looking Ahead—Formulating New Problems
46. Create new problems by varying a given one.

What Are Some Examples of Problem-Solving Strategies?

Since strategies are a combination of skills, a listing (if it were possible) would be even more cumbersome than the list of skills. Examples of some strategies that might be used in the "Machine Hook-Ups" problem follow:

Strategy 1. *Guess* the input; *check* by computing the output number for your guess; if guess does not give the desired output, note the direction of error; *refine* the guess; compute; continue making refinements until the correct output results.

Strategy 2. *Observe* the *patterns* suggested by the input and output numbers for the *a, b, c* entries in the table; *predict* additional output and input numbers by extending both patterns; *check* the predicted input for the *d* entry by computing.

Strategy 3. *Study* the operations suggested in the machine hook-up; *work backwards* through the machine *using previous knowledge* about inverse operations.

An awareness of the strategies being used to solve a problem is probably the most important step in the development of a pupil's problem-solving abilities.

What is the Instructional Approach Used in PSM?

The content objectives of the lessons are similar to those of most textbooks. The difference is in the approach used. First, a wider variety of problem-solving skills is emphasized in the materials than in most texts. Second, different styles of teaching such as direct instruction, guided discovery, laboratory work, small-group discussions, nondirective instruction, and individual work all have a role to play in problem-solving instruction.

Most texts employ direct instruction almost exclusively, whereas similar lessons in PSM are patterned after a guided discovery approach. Also, an attempt is made in the materials to use intuitive approaches extensively before teaching formal algorithms. Each of the following is an integral part of the instructional approach to problem solving.

A. TEACH PROBLEM-SOLVING SKILLS DIRECTLY

Problem-solving skills such as "follow directions," "listen," and "correct errors" are skills teachers expect pupils to master. Yet, such skills as "guess and check," "make a systematic list," "look for a pattern," or "change a problem into one you can solve" are seldom made the object of direct instruction. These skills, as well as many more, need emphasis. Detailed examples for teaching these skills early in the school year are given in the commentaries to the *Getting Started* activities.

B. INCORPORATE A PROBLEM-SOLVING APPROACH WHEN TEACHING TOPICS IN THE COURSE OF STUDY

Drill and practice activities. Each PSM book includes many pages of drill and practice at the problem-solving level. These pages, along with the *Getting Started* section, are easy for pupils and teachers to get into and should be started early in the school year.

Laboratory activities and investigations involving mathematical applications and readiness activities. Readiness activities from such mathematical strands as geometry, number theory, and probability are included in each book. For example, area explorations are used in grade 4 as the initial stage in the teaching of the multiplication and division algorithms and fraction concepts.

Teaching mathematical concepts, generalizations, and processes. Each book includes two or more sections on grade-level content topics. For the most part, these topics are developmental in nature and usually need to be supplemented with practice pages selected from a textbook.

C. PROVIDE MANY OPPORTUNITIES FOR PUPILS TO USE THEIR OWN PROBLEM-SOLVING STRATEGIES

One section of each book includes a collection of challenge activities which provide opportunities for emphasizing problem-solving strategies. Generally, instruction should be nondirective, but at times suggestions may need to be given. If possible, these suggestions should be made in the form of alternatives to be explored rather than hints to be followed.

D. CREATE A CLASSROOM ATMOSPHERE IN WHICH OPENNESS AND CREATIVITY CAN OCCUR

Such a classroom climate should develop if the considerations mentioned in A, B, and C are followed. Some specific suggestions to keep in mind as the materials are used are:

- Set an example by solving problems and by sharing these experiences with the pupils.
- Reduce anxiety by encouraging communication and cooperation. On frequent occasions problems might be investigated using a cooperative mode of instruction along with brainstorming sessions.
- Encourage pupils in their efforts to solve a problem by indicating that their strategies are worth trying and by providing them with sufficient time to investigate the problem; stress the value of the procedures pupils use.
- Use pupils' ideas (including their mistakes) in solving problems and developing lessons.
- Ask probing questions which make use of words and phrases such as
  I wonder if . . . .
  Do you suppose that . . . .
  What happens if . . . .
  How could we find out . . . .
  Is it possible that . . . .
- Reinforce the asking of probing questions by pupils as they search for increased understanding. Pupils seldom are skilled at seeking probing questions but they can be taught to do so. If instruction is successful, questions of the type, "What should I do now?," will be addressed to themselves rather than to the teacher.

What Are the Parts of Each PSM Book?

PROBLEM SOLVING IN MATHEMATICS

Grade 4	Grade 5	Grade 6	Grade 7	Grade 8	Grade 9
Getting Started	Getting Started	Getting Started	Getting Started	Getting Started	Getting Started
Place Value Drill and Practice	Whole Number Drill and Practice	Drill and Practice	Drill and Practice- Whole Numbers	Drill and Practice	Algebraic Concepts and Patterns
Whole Number Drill and Practice	Story Problems	Story Problems	Drill and Practice- Fractions	Variation	Algebraic Explanations
Multiplication and Division Concepts	Fractions	Fractions	Drill and Practice- Decimals	Integer Sense	Equation Solving
Fraction Concepts	Geometry	Geometry	Percent Sense	Equation Solving	Word Problems
Two-digit Multiplication	Decimals	Decimals	Factors, Multiples, and Primes	Protractor Experiments	Binomials
Geometry	Probability	Probability	Measurement-Volume, Area, Perimeter	Investigations in Geometry	Graphs and Equations
Rectangles and Division	Estimation with Calculators	Challenges	Probability	Calculator	Graph Investigations
Challenges	Challenges		Challenges	Percent Estimation	Systems of Linear Equations
				Probability	Challenges
				Challenges	

Notice that the above chart is only a scope of PSM—not a scope and sequence. In general, no sequence of topics is suggested with the exceptions that *Getting Started* activities must come early in the school year and *Challenge* activities are usually deferred until later in the year.

Getting Started Several problem-solving skills are presented in the *Getting Started* section of each grade level. Hopefully, by concentrating on these skills during the first few weeks of school pupils will have confidence in applying them to problems that occur later on. In presenting these skills, a direct mode of instruction is recommended. Since the emphasis needs to be on the problem-solving skill used to find the solution, about ten to twelve minutes per day are needed to present a problem.

Drill And Practice No sequence is implied by the order of activities included in these sections. They can be used throughout the year but are especially appropriate near the beginning of the year when the initial chapters in the textbook emphasize review. Most of the activities are not intended to develop any particular concept. Rather, they are drill and practice lessons with a problem-solving flavor.

Challenges Fifteen or more challenge problems are included in each book. In general, these should be used only after *Getting Started* activities have been completed and pupils have had some successful problem-solving experiences.

Many of the other sections in PSM are intended to focus on particular grade-level content. The purpose is to provide intuitive background for certain topics. A more extensive textbook treatment usually will need to follow the intuitive development.

Teacher Commentaries Each section of a PSM book has an overview teacher commentary. The overview commentary usually includes some philosophy and some suggestions as to how the activities within the section should be used. Also, every pupil page in PSM has a teacher commentary on the back of the lesson. Included here are mathematics teaching objectives, problem-solving skills pupils might use, materials needed, comments and suggestions, and answers.

How Often Should Instruction Be Focused on Problem Solving?

Some class time should be given to problem solving nearly every day. On some days an entire class period might be spent on problem-solving activities; on others, only 8 to 10 minutes. Not all the activities need to be selected from PSM. Your textbook may contain ideas. Certainly you can create some of your own. Many companies now have published excellent materials which can be used as sources for problem-solving ideas. Frequently, short periods of time should be used for identifying and comparing problem-solving skills and strategies used in solving problems.

How Can I Use These Materials When I Can't Even Finish What's in the Regular Textbook?

This is a common concern. But PSM is not intended to be an "add-on" program. Instead, much of PSM can replace material in the textbook. Correlation charts can be made suggesting how PSM can be integrated into the course of study or with the adopted text. Also, certain textbook companies have correlated their tests with the PSM materials.

Can the Materials Be Duplicated?

The pupil lessons may be copied for students. Each pupil lesson may be used as an overhead projector transparency master or as a blackline duplicator master. Sometimes the teacher may want to project one problem at a time for pupils to focus their attentions on. Other times, the teacher might want to duplicate a lesson for individual or small group work. Permission to duplicate pupil lesson pages for classroom use is given by the publisher.

How Can a Teacher Tell Whether Pupils Are Developing and Extending Their Problem-Solving Abilities?

Presently, reliable paper and pencil tests for measuring problem-solving abilities are not available. Teachers, however, can detect problem-solving growth by observing such pupil behaviors as

- identifying the problem-solving skills being used.

- giving accounts of successful strategies used in working on problems.
- insisting on understanding the topics being studied.
- persisting while solving difficult problems.
- working with others to solve problems.
- bringing in problems for class members and teachers to solve.
- inventing new problems by changing problems previously solved.

What Evidence Is There of the Effectiveness of PSM?

Although no carefully controlled longitudinal study has been made, evaluation studies do indicate that pupils, teachers, and parents like the materials. Scores on standardized mathematics achievement tests show that pupils are registering greater gains than expected on all parts of the test, including computation. Significant gains were made on special problem-solving skills tests which were given at the beginning and end of a school year.

Also, when selected materials were used exclusively over a period of several weeks with 6th-grade classes, significant gains were made on the word-problem portion of the standardized test. In general, the greater gains occurred in those classrooms where the materials were used as specified in the teacher commentaries and in-service materials.

Teachers have indicated that problem-solving skills such as *look for a pattern, eliminate possibilities*, and *guess and check* do carry over to other subjects such as Social Studies, Language Arts, and Science. Also, the materials seem to be working with many pupils who have not been especially successful in mathematics. And finally, many teachers report that PSM has caused them to make changes in their teaching style.

Why Is It Best to Have Whole-Staff Commitment?

Improving pupils' abilities to solve problems is not a short-range goal. In general, efforts must be made over a long period of time if permanent changes are to result. Ideally, then, the teaching staff for at least three successive grade levels should commit themselves to using PSM with their pupils. Also, if others are involved, this will allow for opportunities to plan together and to share experiences.

How Much In-Service Is Needed?

A teacher who understands the meaning of problem solving and is comfortable with the different styles of teaching it requires could get by with self in-service by carefully studying the section and page commentaries in a grade-level book. The different styles of teaching required include direct instruction, guided discovery, laboratory work, small group instruction, individual work, and non-directive instructions. The teacher would find the audio tapes for each book and the *In-Service Guide* a valuable resource and even a time saver.

If a school staff decides to emphasize problem solving in all grade levels where PSM books are available, in-service sessions should be led by someone who has used the materials in the intended way. For more information on this in-service see the *In-Service Guide*.

What Materials Are Needed?

PROBLEM—SOLVING PROGRAM

REQUIRED MATERIALS	Grade 4	5	6	7	8	9
blank cards	X	X	X	X	X	X
bottle caps or markers	X			X		
calendar						X
calculators (optional for some activities)	X	X	X	X	X	X
cm squared paper, strips and singles						X
coins				X	X	
colored construction paper (circle fractions)	X	X				
cubes		X	X	X		X
cubes with red, yellow and green faces					X	
Cuisenaire rods (orange and white) or strips of paper		X				
dice (blank wooden or foam, for special dice)	X					
dice, regular (average 2 per student)	X	X	X	X	X	
geoboards, rubber bands, and record paper	X		X			
graph paper or cm squared paper				X		X
grid paper (1")			X			
metric rulers		X			X	X
phone books, newspapers, magazines		X		X		
protractors and compasses					X	
scissors	X	X	X	X		
spinners (2 teacher-made)			X			
tangrams	X					
tape measures				X		
thumbtacks (10 per pair of students)		X				
tile	X		X			
tongue depressors	X					
uncooked spaghetti or paper strips			X			

PSM   Rev. 1982

RECOMMENDED MATERIALS	Grade 4	5	6	7	8	9
adding machine tape				X		
centimetre rulers			X	X		
colored pens, pencils, or crayons		X				
coins, toy or real	X					
coins (two and one-half)						X
cubes					X	
demonstration ruler for overhead		X				
dominoes					X	
geoboard, transparent (for overhead)	X		X			
money - 20 $1.00 bills per student			X			
moveable markers		X	X		X	
octahedral die for extension activity				X		
overhead projector	X	X	X	X	X	X
place value frame and markers			X			
straws, uncooked spaghetti, or toothpicks		X				
transparent circle fractions for overhead	X	X				

PSM   Rev. 1982

Where Can I Find Other Problem Solving Materials?

RESOURCE BIBLIOGRAPHY

The number in parentheses refers to the list of publishers on the next page.

For <u>students</u> <u>and</u> <u>teachers</u>:

AFTER MATH, BOOKS I—IV by Dale Seymour, et al.
    Puzzles to solve -- some of them non-mathematical. (1)

AHA, INSIGHT by Martin Gardner
    Puzzles to solve -- many of them non-mathematical. (3)

THE BOOK OF THINK by Marilyn Burns
    Situations leading to a problem-solving investigation. (1)

CALCULATOR ACTIVITIES FOR THE CLASSROOM, BOOKS 1 & 2 by George Immerzeel and
    Earl Ockenga
    Calculator activities using problem solving. (1)

GEOMETRY AND VISUALIZATION by Mathematics Resource Project
    Resource materials for geometry. (1)

GOOD TIMES MATH EVENT BOOK by Marilyn Burns
    Situations leading to a problem-solving investigation. (1)

FAVORITE PROBLEMS by Dale Seymour
    Problem solving challenges for grades 5-7. (3)

FUNTASTIC CALCULATOR MATH by Edward Beardslee
    Calculator activities using problem solving. (4)

I HATE MATHEMATICS! BOOK by Marilyn Burns
    Situations leading to a problem solving investigation. (3)

MATHEMATICS IN SCIENCE AND SOCIETY by Mathematics Resource Project
    Resource activities in the fields of astronomy, biology, environment,
    music, physics, and sports. (1)

MIND BENDERS by Anita Harnadek
    Logic problems to develop deductive thinking skills. Books A-1, A-2, A-3,
    and A-4 are easy. Books B-1, B-2, B-3, and B-4 are of medium difficulty.
    Books C-1, C-2, and C-3 are difficult. (6)

NUMBER NUTZ (Books A, B, C, D) by Arthur Wiebe
    Drill and practice activities at the problem solving level. (2)

NUMBER SENSE AND ARITHMETIC SKILLS by Mathematics Resource Project
    Resource materials for place value, whole numbers, fractions, and decimals. (1)

The <u>Oregon</u> <u>Mathematics</u> <u>Teacher</u> (magazine)
    Situations leading to a problem solving investigation. (8)

PROBLEM OF THE WEEK by Lyle Fisher and William Medigovich
    Problem solving challenges for grades 7-12.   (3)

RATIO, PROPORTION AND SCALING by Mathematics Resource Project
    Resource materials for ratio, proportion, percent, and scale drawings.   (1)

STATISTICS AND INFORMATION ORGANIZATION by Mathematics Resource Project
    Resource materials for statistics and probability.   (1)

SUPER PROBLEMS by Lyle Fisher
    Problem solving challenges for grades 7-9.   (3)

For teachers only:

DIDACTICS AND MATHEMATICS by Mathematics Resource Project   (1)

HOW TO SOLVE IT by George Polya   (3)

MATH IN OREGON SCHOOLS by the Oregon Department of Education   (9)

PROBLEM SOLVING: A BASIC MATHEMATICS GOAL by the Ohio Department of Education   (3)

PROBLEM SOLVING: A HANDBOOK FOR TEACHERS by Stephen Krulik and Jesse Rudnik   (1)

PROBLEM SOLVING IN SCHOOL MATHEMATICS by NCTM   (7)

Publisher's List

1. Creative Publications, 3977 E Bayshore Rd, PO Box 10328, Palo Alto, CA 94303

2. Creative Teaching Associates, PO Box 7714, Fresno, CA 93727

3. Dale Seymour Publications, PO Box 10888, Palo Alto, CA 94303

4. Enrich, Inc., 760 Kifer Rd, Sunnyvale, CA 94086

5. W. H. Freeman and Co., 660 Market St, San Francisco, CA 94104

6. Midwest Publications, PO Box 448, Pacific Grove, CA 93950

7. National Council of Teachers of Mathematics, 1906 Association Dr, Reston, VA
    22091

8. Oregon Council of Teachers of Mathematics, Clackamas High School,
    13801 SE Webster St, Milwaukie, OR 97222

9. Oregon Department of Education, 700 Pringle Parkway SE, Salem, OR 97310

Grade 6

I. GETTING STARTED

I. GETTING STARTED

Teachers usually are successful at
teaching skills in mathematics. Besides
computation skills, they emphasize skills
in following directions, listening, de-
tecting errors, explaining, recording,
comparing, measuring, sharing, ...
They (You!) can also teach problem-
solving skills. This section is designed
to help teachers teach and students learn
specific problem-solving skills.

10 ? 30? 20 ? 25 ? 26

Some Problem-Solving Skills

Five common but powerful problem-solving skills are introduced in this
section. They are:

. guess and check
. look for a pattern
. make a systematic list
. make and use a drawing or model
. eliminate possibilities

Students <u>might</u> use other skills to solve the problems. They can be
praised for their insight but it is usually a good idea to limit the list
of skills taught during the first few lessons. More problem-solving skills
will occur in the other sections.

An Important DON'T

When you read the episodes that follow in this <u>Getting Started</u> section
notice how the lessons are <u>very teacher directed</u>. The main purpose is to
teach the problem-solving skills. Teachers should stress the skills
verbally and write them on the board. <u>Don't</u> just ditto these activities
and hand them out to be worked. Teacher direction through questions,
summaries, praise, etc. is <u>most</u> important for teaching the problem-solving
skills in this section. We want students to focus on specific skills
which will be used often in all the sections. Later, in the <u>Challenge
Problems</u> section, students will be working more independently.

Using The Activities

If you heed the important <u>Don't</u> on the previous page, you are on your way to success! The problems here should fit right in with your required course of study as they use whole number skills, elementary geometry and money concepts. In most cases, students will have the prerequisites for the problems in this section although you might want to check over each problem to be sure.

No special materials are required although markers, coins, cubes and toothpicks are helpful for some of the problems. The large type used for the problems makes them easier to read if they are shown on an overhead screen. In most cases students can easily copy the problem from the overhead. At other times you might copy the problem onto the chalkboard.

When And How Many

The <u>Getting Started</u> section should be used at the beginning of the year as it builds background in problem-solving skills for the other sections. As the format indicates <u>only one problem per day</u> should be used. Each should take less than twelve minutes of classtime if the direct mode of instruction is used. The remainder of the period is used for a lesson from the textbook or perhaps an activity from the <u>Drill and Practice</u> section of these materials.

REMEMBER: One Problem Per Day when you are using this <u>Getting Started</u> section.

Guess And Check

The episode that follows shows how one teacher teaches the skill of guess and check. Notice that she very closely directs the instruction and constantly uses the terminology.

It is near the beginning of the year and Ms. Jones is about to start a math lesson. After getting the attention of the class, she begins.

Ms. J: I'm trying to figure out a number. Twice the number minus 18 equals 34. I wonder if the number is 10. Is it? (Waits for hands.) Tim?

Tim: Nope!

Ms. J: How do you know?

Tim: 'Cause twice 10 is 20, minus 18 gives 2.

Ms. J: By checking my guess you found out it was off. Well, is the number 30? Mary?

Mary: (Thinking out loud.) Twice 30 is 60, minus 18 is 42. No--too big.

Ms. J: Thirty is too big? What can you say about 10?

Tim: It was too small.

Ms. J: Guessing and checking helped you decide 10 is too small and 30 is too big. Let's refine the guesses. Refine means to make better guesses. Can you make a better guess? (Some puzzled, some thinking, some hands.) Larry?

Larry: Try 20, it's in the middle.

Sue: (Computing on her paper.) No, 20 is too small, it gives 22.

Ms. J: Does anyone think they know the answer? Scott?

Scott: 25

Ms. J: Why did you pick 25?

Scott: 20 is too small and 30 is too big--Let's see--twice 25 is 50, minus 18 is 32. Oh! It's got to be 26!

Ms. J: Scott, you made a close guess, checked it and refined it to get the right answer. Did you know that guess, check and refine is a good way to solve problems? We're going to use it a lot this year. I'm putting it up on the wall so we'll all remember how important it is! Let's try another problem...

GUESS AND CHECK

WEEK 1 - DAY 1

a. I'm thinking of a number. Twice the number minus 18 equals 34. What is the number?

b. I'm thinking of a number. Three times the number plus 23 equals 53. What is the number?

c. I'm thinking of a number. Multiply the number by itself and then by itself again and you will get 216. What is the number?

****

WEEK 1 - DAY 2

Terry Cloth has 58¢ consisting of 9 coins. She does not have a half-dollar. What are the coins?

****

WEEK 1 - DAY 3

Find a path from start to finish with a product of 630. You may only go through the open gates.

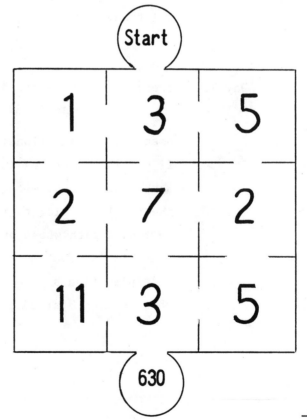

Guess And Check

Day 1. Answers: a.   26   b.   10   c.   6

          Comments and suggestions:

       . An outline of how to introduce these problems is given in
         Ms. Jones' lesson on page 3 of the overview to this section.
         Remember, these are meant to be teacher directed.

       . You can include division by using the problem below.

          d.   I'm thinking of a number.   Divide the number by 3.
             Then subtract 5 and the answer is 17.   What is the
             number?   - answer, 66.

Day 2. Answers: She has 5 dimes, 1 nickel, and 3 pennies.   Or the answer
               could be 1 quarter, 1 dime, 4 nickels, and 3 pennies..

          Comments and suggestions:

       . Pupil guesses should focus on 6 coins to make 55¢ since Terry
         must have 3 pennies and can't have 8 pennies (unless she is
         allowed a half-dollar).

       . Actual use of money or play coins may help in obtaining the
         solution.

Day 3. Answers: The paths are 3, 5 (top), 2, 7, 3   or   3, 7, 2,
               5 (bottom), 3.

          Comments and suggestions:

       . If pupils do not realize that every path must pass through
         both 3's, discuss with them how this knowledge would make
         the problem easier to refine--3 x 3 is 9 so the problem is
         to find a path whose product is 70, since 9 x 70 = 630.

       . "Big Tic Tac Toe" on page 59 emphasizes guess and check and
         multiplication skills.   The activity is a good following to
         this Getting Started problem.

Guess And Check (cont.)

WEEK 1 - DAY 4

Place the numbers
1, 4, 6, 7, 8, 9,
and 10 in the circles
to make each side of
the pentagon add to 19.

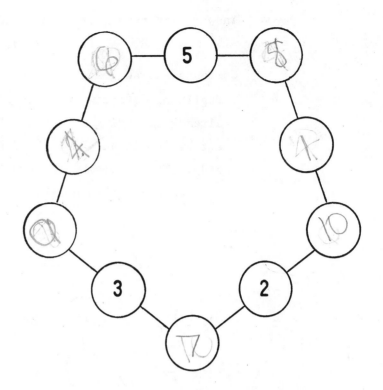

****

WEEK 1 - DAY 5

Make this drawing.
Place the numbers
1, 2, 3, 4, 5, 6,
and 7 in the circles.
Make the sum along
each line equal to 13.

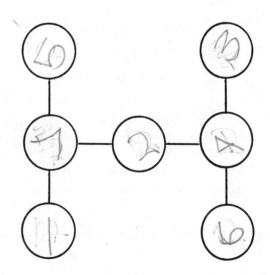

Guess And Check

Day 4. Answer: Starting with the 5 and moving clockwise--

5, 8, 1, 10, 2, 7, 3, 9, 4, 6

Comments and suggestions:

. Pupils may start with guesses along any side of the pentagon already including a number. For example, using the 5 side, either the 10 and 4 or the 6 and 8 can be used along that side. The problem is solved by trying these possible combinations and then checking the remaining sides.

Day 5. Answer:

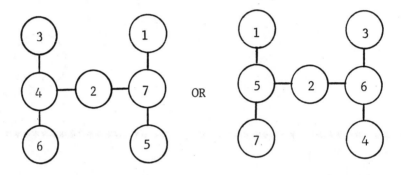

Comments and suggestions:

. Use of markers labeled with the digits will allow pupils to try many guesses without having to erase mistakes.

. If both solutions are found, a discussion of what the solutions have in common will show that the two circles used in more than one row must equal 11. The only possibilities are 4 + 7 and 5 + 6.

Look For A Pattern

By now your pupils are familiar with the skill <u>guess</u> and <u>check</u>. One teacher introduced the next problem-solving skill, <u>look</u> <u>for</u> <u>a</u> <u>pattern</u> in this way.

Mr. Todd: Who remembers what method we used to solve problems last week?

Teri: We guessed.

Mr. Todd: Is that all?

Sam: Guess, check and refine! It's up on the poster!

Mr. Todd: That's right and we're going to add another problem-solving skill to the poster today. (Writes it up.) What does it say?

Class: Look for a pattern.

Mr. Todd: Look for a pattern--this week we are going to practice looking for patterns. Here's our problem. (Shows 1, 4, 7, 10, 13, __, __, __ on the overhead.) We want to fill in the next three blanks. Can you see a pattern in the numbers?

Sid: They go up by 3.

Mr. T: What goes in the blanks?

Several: 16, 19, 22

Mr. T: Let's try the next one. (Shows 29, 28, 26, 23, 19, __, __, __) Who can tell me the numbers and the pattern they used?

Mary: 14, 8, 1. They go down by 1, then by 2, then by 3, and so on.

Mr. T: Good. Sue, you are frowning, what's wrong?

Sue: I saw odd, even, odd, even, but it didn't help to get the numbers.

Mr. T: Well, it is a good pattern though. That's pretty neat to remember about odd and even. Sometimes one pattern isn't enough to solve a problem and we have to look for another one. Let's see if we can solve the rest ...

LOOK FOR A PATTERN

WEEK 2 - DAY 1

Continue these patterns. Fill in the next three blanks.

a.   1, 4, 7, 10, 13, _16_ , _19_ , _22_

b.   29, 28, 26, 23, 19, _14_ , _8_ , _1_

c.   2, 6, 18, 54, ___, ___, ___

****

WEEK 2 - DAY 2

Draw the next <u>two</u> figures to continue this pattern.

a.   . :   . : :   . : : :   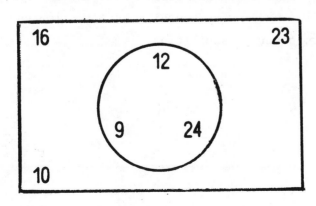 , _____

b.   How many dots in each figure? _3_ , _5_ , _7_ , _9_ , _11_

****

WEEK 2 - DAY 3

Study the diagram. Place the numbers from 8 to 26 either inside or outside the circle according to the pattern shown.

```
 ┌──────────────────────────────┐
 │ 16                        23  │
 │          ⟋‾‾‾⟍              │
 │        ⟋   12   ⟍           │
 │       │          │          │
 │       │          │          │
 │        ⟍  9   24⟋           │
 │          ⟍___⟋              │
 │ 10                            │
 └──────────────────────────────┘
```

-13-

Look For A Pattern

Day 1.   Answers: a. 16, 19, 22   b. 14, 8, 1   c. 162, 486, 1458

Comments and suggestions:
. Pupils usually like to look for patterns like these. Pupils
  may find patterns other than the ones shown. If a good argu-
  ment is provided to support their answers, they should be
  accepted.

Day 2.   Answers: • • • •   • • • • •   b. 3, 5, 7, 9, 11
                   • • • • • , • • • • • •

Comments and suggestions:
. Discuss patterns pupils see. Some will see one dot followed
  by 2, 4, 6, ... dots. Others will see a certain number of
  dots on the bottom row, then one less dot on the top row.
  Others will see--??
. "Surface Area Patterns With Cubes" on page 165 emphasizes
  looking for patterns within cube arrangements. The activity
  is a good follow-up to this Getting Started problem.

Day 3.   Answers: Inside the circle   9, 12, 15, 18, 21, 24
                   Outside the circle { 8, 10, 11, 13, 14, 16, 17, 19
                                        20, 22, 23, 25, 26

The numbers inside the circle are multiples of 3.

Comments and suggestions:
. Pupils may find other patterns. Some may see these numbers
  as divisible by 3 or as counting by 3. Allow pupils to
  brainstorm other possibilities.

Look For A Pattern (cont.)

WEEK 2 - DAY 4

Find these answers:

a.   37 x 3 = _____

b.   37 x 6 = _____

c.   37 x 9 = _____

d. Predict the answer for 37 x 15. _____ Now check it.

e. Predict the multipliers for 37 x _____ = 444

f.   and 37 x _____ = 999.

****

WEEK 2 - DAY 5

a. Two points can be connected with
   1 line segment.      1

b. Three points can be connected with
   3 line segments.      3

c. Four points can be connected with
   6 line segments.      6

d. Five points can be connected with
   _____ line segments.      _____

e. Predict, then check, the number of
   line segments that will connect six points.   _____

f. Predict, then check, the number of line
   segments that will connect ten points.   _____

<u>Look</u> <u>For</u> <u>A</u> <u>Pattern</u>

Day 4. Answers: a. 111   b. 222   c. 333

       d. 37 x 15 = 555   e. 37 x 12 = 444   f. 37 x 27 = 999

Comments and suggestions:

. Organizing the problems in a table may help pupils find the
 pattern and then predict what will happen with other multiples
 of 3.

. Perhaps use of a calculator will increase the motivation level
 for the problem.

Day 5. Answers: d. 10 line segments   e. 15 line segments

         f. 45 line segments

Comments and suggestions:

. Some pupils will see this as a geometric problem while others
 will see it as an arithmetic problem. Actually drawing the
 diagrams is appropriate for the former while the latter will
 see a pattern in the numbers as shown below.

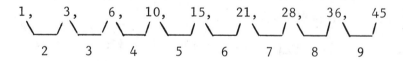

Make A Systematic List

The skills of <u>guess</u> and <u>check</u> and <u>look</u> <u>for</u> <u>a</u> <u>pattern</u> are often used with the skill of <u>make</u> <u>a</u> <u>systematic</u> <u>list</u>. Use of the list helps pupils be certain that all possibilities have been tried and all solutions have been found. This skill is often difficult for pupils and much guidance in setting up the lists is needed. One teacher introduced the skill of <u>make</u> <u>a</u> <u>systematic</u> <u>list</u> with this problem.

> Carla and Nick were playing a factors game. Carla would name a factor of 48 and Nick would give the other factor. Show the ways Carla and Nick could play the game.

Mr. Benton: What are some of the ways Carla and Nick can play the game? Scott? 6 and 8. Good. Sue? 4 and 12. Joe? 2 and 24. Any more? (Silence.) Do we have them all? Let's <u>make</u> <u>a</u> <u>systematic</u> <u>list</u>. (Writes the skill on the poster.) Which pair should we write first?

Ted: 2 and 24 because 2 is the smallest, then 4 and 12, then 6 and 8.

Mr. B: You know, I can't tell which number was chosen by Carla. Maybe we should label our list. Now did we miss any?

Carla	Nick
2	24
4	12
6	8

Mary: Well, we could switch them--you know, Carla 24 and Nick 2.

Mr. B: That gives us six solutions. Could Carla's numbers get smaller?

Jim: No -- Oh, yes! 1 and 48 and 48 and 1.

Dick: There's more. How about 3 and 16 and 16 and 3.

Mr. B: Great. Do you see how <u>making</u> <u>a</u> <u>systematic</u> <u>list</u> can help solve a problem? Let's try another ...

MAKE A SYSTEMATIC LIST

WEEK 3 - DAY 1

Carla and Nick were playing a factors game. Carla would name a
factor of 48 and Nick would give the other factor. Make a table
to show the different ways Carla and Nick could play the 48 game.

****

WEEK 3 - DAY 2

What scores are possible if three
darts are thrown at this dart board?
Assume that none of the darts miss
the target. A table is started
for you.

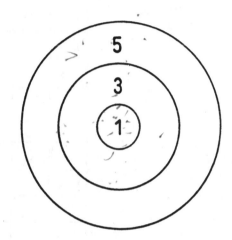

1	3	5	Total
XXX			3
XX	X		5

****

WEEK 3 - DAY 3

Dee Phie and Ossi Phie were playing a dice game. They needed to
know all the possible products they could get by rolling two dice.
What products are possible?

Make A Systematic List

Day 1. Answers:

Carla's factor	1	2	3	4	6	8	12	16	24	48
Nick's factor	48	24	16	12	8	6	4	3	2	1

Comments and suggestions:

. Pupils will usually begin by listing the factors at random. Systematically listing the factors in a table usually convinces pupils that all pairs have been found.

. Pupils should see that as Carla's factors increase, Nick's factors decrease.

Day 2. Answers:

1	3	5	Total
xxx			3
xx	x		5
xx		x	7
x	xx		7
x	x	x	9
x		xx	11
	xxx		9
	xx	x	11
	x	xx	13
		xxx	15

Comments and suggestions:

. Emphasize the scheme used in the list. First combinations with 3 darts in the 1-region, then 2 darts, then 1 dart, and then 0 darts in the 1-region. Of course the list could be generated using the 5-region or the 3-region.

. "Exchanges" on page 55 emphasizes making a systematic list in the extension to the problem. Dealing with place value, the activity makes a good follow-up to this Getting Started problem.

Day 3. Answers:

X	1	2	3	4	5	6
1	1	2	3	4	5	6
2	2	4	6	8	10	12
3	3	6	9	12	15	18
4	4	8	12	16	20	24
5	5	10	15	20	25	30
6	6	12	18	24	30	36

Comments and suggestions:

. A multiplication matrix is one way of showing the products. Some pupils may prefer a list as shown below.

1x1	2x1	3x1	4x1	5x1	6x1
1x2	2x2	3x2	4x2	5x2	6x2
1x3	2x3	3x3	4x3	5x3	6x3
1x4	2x4	3x4	4x4	5x4	6x4
1x5	2x5	3x5	4x5	5x5	6x5
1x6	2x6	3x6	4x6	5x6	6x6

Make A Systematic List (cont.)

WEEK 3 - Day 4

Mr. Jones works in a shop that produces 4-legged tables and 3-legged stools. Nine customers ordered 8 items each. Each order was different. How many legs are needed for each of the customers?

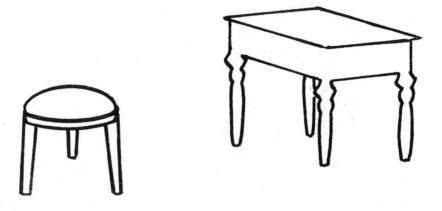

****

WEEK 3 - DAY 5

Sam has three markers--one blue, one green, and one yellow. If she arranges them in a row, show the different arrangements she could make.

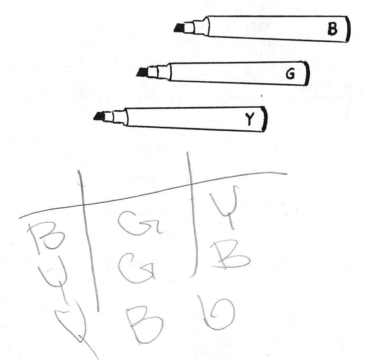

Day 4. Answers:

Tables	Stools	Legs
8	0	32
7	1	31
6	2	30
5	3	29
4	4	28
3	5	27
2	6	26
1	7	25
0	8	24

Comments and suggestions:

. The systematic lists shows a **pattern** so pupils can be reasonably sure all possibilities have been considered. As the number of tables goes from 8 to 0, the number of stools goes from 0 to 8. The number of legs shows a decrease of 1 each time as a 4-legged table is replaced by a 3-legged stool.

Day 5. Answers: BGY,   BYG,   GBY,   GYB,   YBG,   YGB

Comments and suggestions:

. Encourage the use of some kind of "abbreviation" for making the list. Actual colored markers would be effective to show the arrangements.
. A tree diagram (shown below) makes another visual display of the problem.

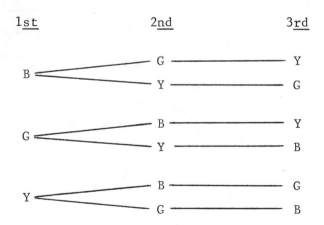

-22-

Make And Use A Drawing Or Model

This problem was used to illustrate the skill of make and use a drawing.

In a make-believe horse race between five famous horses,

    a. Citation finished 1 length ahead of Seattle Slew.

    b. Spectacular Bid finished ahead of Citation but behind Secretariat.

    c. Man-O-War finished 4 lengths ahead of Seattle Slew and 1 length behind Spectacular Bid.

What is the finish place of each horse?

Pupils were allowed to think about the problem for a few minutes. Some reasoned out the facts but others weren't sure how to proceed.

Mrs. Thompson: Nora, you wrote something about the problem. Will you share it with us?

Nora: I read the clues and guessed that Secretariat finished first. I think it is correct but I'm not sure.

Mrs. T: That's good, Nora. Did anyone try something else?

Tom: I made a line for the race track and tried to place each horse along the line.

Mrs. T: Excellent, Tom. You have made a drawing of the problem. Perhaps it might help to divide your line to show lengths ahead or behind.

Tom: Yes, that will help.

Juanita: I've got it! By making a drawing clue c is the one that really helps. Nora was right; Secretariat was first and Seattle Slew was last.

Mrs. T: Very nice, Juanita. Making a drawing of a problem can be very helpful, can't it?

MAKE AND USE A DRAWING OR MODEL

WEEK 4 - DAY 1

In a make-believe horse race between five famous horses,

a. Citation finished 1 length ahead of Seattle Slew.

b. Spectacular Bid finished ahead of Citation but behind
    Secretariat.

c. Man-O-War finished 4 lengths ahead of Seattle Slew and
    1 length behind Spectacular Bid.

What was the finish place of each horse?

****

WEEK 4 - DAY 2

This cube has the letters M, N, O, P, Q, and R on it.
Three views of the cube are shown. Complete the other three views.

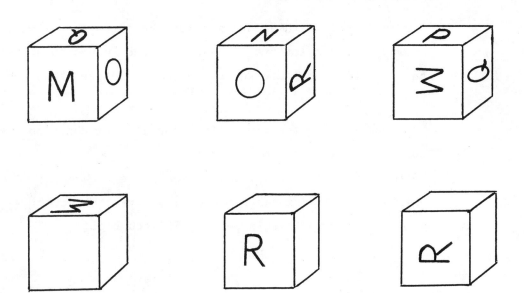

Day 1. Answers: 1st-Secretariat, 2nd-Spectacular Bid, 3rd-Man-O-War,
                   4th-Citation,   5th-Seattle Slew

Comments and suggestions:

. A line marked in lengths is effective for showing this problem.

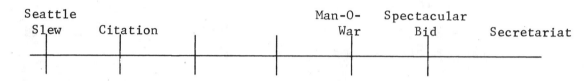

Seattle
Slew   Citation   Man-O-   Spectacular
                                                 War   Bid   Secretariat

. You might ask pupils how many lengths Secretariat won by. The
  information given doesn't allow pupils to figure this out.

Day 2. Answers:

Comments and suggestions:

. Provide pupils with a cube and let them draw the views that are
  shown. All six letters will show on the cube. Then pupils need
  only position the cube so the indicated letter shows and draw
  the other letters.

. Some pupils will see clues like "the stem of the P points towards
  the left side of the M."

. "Helping Out" on page 65 emphasizes using a physical model.
  Dealing with fractional parts of a square, the activity makes
  a good follow-up to this Getting Started problem.

Make And Use A Drawing Or Model (cont.)

WEEK 4 - DAY 3

Use four tiles or squares. Make all the arrangements you can that place the four squares side by side.

O.K.   Not O.K.

****

WEEK 4 - DAY 4

Three cannibals are on one side of a river and three missionaries are on the other side. Both groups want to get to the other side. The cannibals have a boat which holds exactly two people. How can the groups get across the river?   (Oh yes, the number of cannibals on either side can never be more than the number of missionaries.)

Day 3. Answers: Five arrangements are possible.

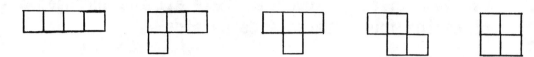

Comments and suggestions:

. Pupils will think many other arrangements are possible but
turning or flipping a possible arrangement should duplicate
one of the above answers.

Day 4. Answers: Five trips are needed.

Start	CCC		MMM
Trip			
1	CC	---C-->	MMM
2	CC	<--MM--	M C
3	MC	--MC-->	M C
4	MC	<--MM--	CC
5	MMM	--C-->	CC

Comments and suggestions:

. Pupils could use markers to move forth and back over an
imaginary river. Or six pupils could actually walk through
the situation. In either case, record keeping is important
to avoid repeating false starts.

Make And Use A Drawing Or Model (cont.)

WEEK 4 - DAY 5

Get 12 toothpicks (or straws) all the same length.

a. Use all 12 to make six triangles. Sketch your solution.

b. Use 9 toothpicks to make five triangles.

Day 5. Answers:

a. 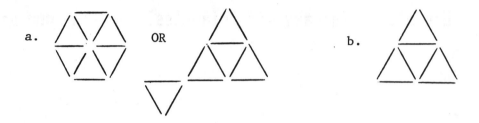 OR   b.

Comments and suggestions:

. One pupil found the second answer to part (a) by actually
working part (b) first. He found 5 triangles using 9
toothpicks and just "tacked on" the other 3 toothpicks to
make a sixth triangle.

. Some pupils will need a hint to break the mindset that all
triangles need to be the same size.

Eliminate Possibilities

    Usually pupils try to solve a problem by looking directly for the answer, but sometimes it is more helpful to identify or list possible answers and then eliminate incorrect answers. This narrows the search for a correct answer and in some cases leaves only the correct answer.

    Ms. Strate used this problem to introduce the skill eliminate possibilities.

> When Tammy puts her marbles in groups of 5, she has 1 marble left over.
>
> When she puts her marbles in groups of 6, she has 1 marble left over.
>
> She has less than 40 marbles.
>
> How many does she have?

Ms. S: Let's list the possible answers using only the first piece of information. If the marbles are in groups of 5 with one left over, how many might she have?

Lorne: 6, 11, 16, ... just add one to multiples of 5.

Ms. S: O.K. Let's list the multiples of 5 first and then add one.
(Writes 5, 10, 15, 20, 25, 30, 35, 40
         6, 11, 16, 21, 26, 31, 36, 41)
Now can we eliminate any of them?

Megan: You can cross off 41. It's too big.

Maureen: Cross off 6 and 36 because there wouldn't be any left if they were put in piles of 6.

Frank: The only one that gives a remainder of 1 is 31.

Teri: What about 1? Doesn't that count?

Ms. S: What do you think? Should we allow 1 as an answer?

Manley: I think so. It's less than 40.

Larry: No, because there wouldn't be any groups of 5 or 6.

Ms. S: The answers depend on how we interpret the problem. I'll leave that up to you as individuals. Do you see how it helped to list the possibilities and then eliminate certain ones? This is our new problem-solving skill: eliminate possibilities.

ELIMINATE POSSIBILITIES

WEEK 5 - DAY 1

When Tammy puts her marbles in groups of 5, she has 1 marble left over.

When she puts her marbles in groups of 6, she has 1 marble left over.

She has less than 40 marbles.

How many does she have?

****

WEEK 5 - DAY 2

Use the digits 2, 3, 7, 8. Make two multiplication problems so each product is larger than 5000.

PSM 81

Eliminate Possibilities

Day 1. Answer: 31 or 1   (See introductory commentary--pupils might
                                not want to allow 1 as an answer.)

Day 2. Answer: 72   73
                            x 83   x 82

Comments and suggestions:

. Pupils can eliminate 2 and 3 as possibilities in the tens
  place by trying 37 x 82 to get 3034 and seeing this is too
  small.

. The above answers could also be written as

                    83   82
                 x 72   x 73

Eliminate Possibilities (cont.)

WEEK 5 - DAY 3

```
  A A A
  B B B
+ C C C
-------
  D D D
```

A, B, C, and D each stand for a
different counting number. Which
of these is a possibility for DDD?

333   444   555   666   777

****

WEEK 5 - DAY 4

I am less than 100.

I am an odd number.

I am a multiple of 5.

I am divisible by 3.

The sum of my digits is an odd number.

Who am I?

Eliminate Possibilities

Day 3. Answers: 666 and 777

Comments and suggestions:

. Since the problem says <u>counting numbers</u>, zero cannot be
   used. So 333 cannot be done as 000 + 111 + 222.

. Possible answers for 666 and 777 are 111 + 222 + 333 = 666
   and 111 + 222 + 444 = 777.

Day 4. Answer: 45

Comments and suggestions:

. Have pupils read the entire problem.
. Which of the following would be best for the pupil to do?
   List all the whole numbers less than 100? List the odds
   less than 100? The multiples of 5? Multiples of 3?
. One possibility is to list the multiples of 5, circle
   those divisible by 3, "X out" the circled even numbers,
   then check each circled number until finding an odd sum.

5,   10,   (15,)   20,   25,   (30,)   35,   40,   (45,)   50,
          No   Yes

55,   (60,)   65,   70,   (75,)   80,   85,   (90,)   95
                          No

-36-

5. Ellen, Sam, June, and Jose are seated around a table.
   Their jobs are teacher, banker, carpenter, and artist.
   Use these clues to decide which seating arrangement below is
   the correct one.

   1. The artist sat on June's left.
   2. The banker sat across from Sam.
   3. Ellen and Jose sat next to each other.
   4. A woman sat on the carpenter's left.

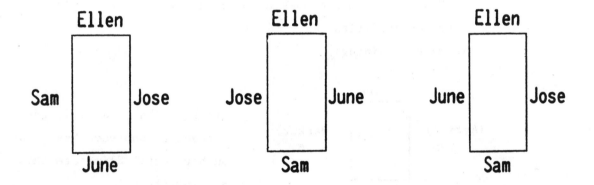

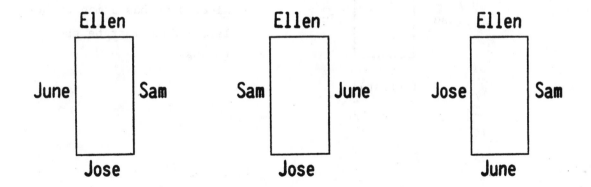

Eliminate Possibilities

Day 5. Answer:

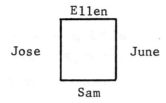

Comments and suggestions:

. Assume that Sam and Jose are men and June and Ellen are women.

. Statement 3 is the easiest to use and will eliminate the lower-left two possibilities.

. For each remaining possibility, start by labeling the person to June's left as artist" and the person across from Sam as "banker." This causes a contradiction in both right-hand possibilities.

. We have remaining:

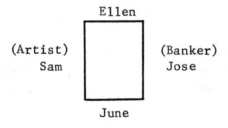

Either Ellen or June is the carpenter, but each has a man on her left. Eliminate this possibility.

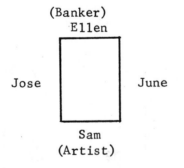

Jose must be the carpenter since Jane has a man on her left. June must be the teacher.

-38-

Grade 6

II. DRILL AND PRACTICE

II. DRILL AND PRACTICE

Most sixth-grade classes are a collection of pupils with varying levels of skills. What can be done to provide additional practice and learning for all pupils? One solution is to offer drill and practice through problem-solving activities. While Tim is remembering that 3 x 8 is 24, not 32, Hosea might be figuring out all the possible ways to

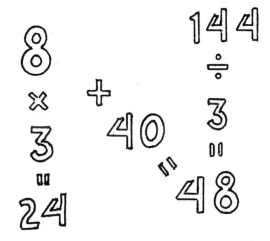

complete _____ x _____ = 24. Another solution is to use games. If the games are combined with problem solving, many objectives are accomplished at the same time: motivation, computation practice, and emphasizing of problem-solving skills.

One marvelous discovery in using problem-solving skills with a mixed class is that some pupils who are unskilled in computation are really good at seeing patterns or figuring out ways to solve problems. Some who hate ordinary drill problems will fill a page with computation to try to solve a problem.

Using The Activities

The activities in this section can be incorporated with the regular teaching and review of place value, whole number, fraction and decimal skills. Realizing that pupils have different skills, many activities have suggestions for simplifying or extending the problems. Many of the activities are meant to be teacher-directed while others can be finished by pupils after a brief introduction.

DIGIT DRAW ACTIVITIES
(Ideas for Teachers)

Ten digit cards marked 0-9 can be used for a variety of activities.

A. Have pupils make the diagram to the right.

B. Suggest a goal, such as, getting the largest possible sum.

C. Shuffle and draw digits, one at a time. Pupils must write each digit, as it is drawn, until all spaces are filled.

D. Compare and discuss results.

This activity is very adaptable. For example, in the 3-digit by 3-digit addition problem above, pupils may not get the largest sum the first time. The teacher can have the pupils use the same six digits to find the largest sum; then use the same six digits to find the smallest sum; and then use those digits to find the sum closest to, say, 700.

These activities can be used for drill and practice, concept development and/or diagnosis. Each is highly individual, as many pupils will have a unique problem. See the next page for other suggested formats. Further variations could include:

- replacing a drawn digit so it can be used again

- restricting certain numbers, e.g., no zero is used with division and fraction problems

- allowing a special reject box giving pupils the chance to discard an unfavorable draw

- adapting whole number activities to decimal activities by inserting a decimal point(s).

PSM 81

Digit Draw Activities

Mathematics teaching objectives:

  . Develop place value concepts.

  . Develop informal probability concepts.

  . Practice computation skills.

Problem-solving skills pupils might use:

  . Break problem into manageable parts.

  . Make decisions based upon data.

  . Recognize limits and/or eliminate possibilities.

Materials needed:

  . Digit cards 0-9.

Comments and suggestions:

  . This activity works well as an opener at the beginning of the class
    or as an ending for that last few minutes of a class when all other
    activities have been completed...See pages 43 and 45 for other sug--
    gestions and many variations.

  . After some trials, pupils can be encouraged to share what strat-
    egies they use. "if a 9 or 8 is drawn first, put it in the 100's
    place. If a little digit comes first, put it in the 1's place.
    It's hard to decide what to do with a 5 or 6. Sometimes it's
    just luck!"

  . The strategies pupils use to solve problems 1, 2 and 3 on the
    second page will vary greatly. In problem 3 one pupil might
    write out lots of possibilities to see which one gives smaller
    answers. This pupil needs encouragement in organizing the trials
    and making conclusions from the trials that give larger answers.
    Another might try to solve the problem by analyzing the sub-
    traction process. Others might take the problem to their parents
    or an older friend for some shared problem solving. Pupils can
    be complimented on their efforts even if they don't find the
    smallest difference. The smallest difference found could be
    posted and changed as someone finds a smaller difference.

Answers:

Answers will vary according to the digits drawn and the places
where pupils put the digits.

Digit Draw Activities (cont.)

OTHER FORMATS

Place Value: ☐,☐ ☐ ☐ largest number, smallest number, or closest to 5000

Ordering: ☐ ☐ < ☐ ☐ < ☐ ☐

Addition:
☐ ☐   ☐ ☐ ☐
☐ ☐ + ☐ ☐ ☐
+ ☐ ☐

Subtraction:
☐ ☐ ☐   ☐ ☐ ☐
− ☐ ☐   − ☐ ☐ ☐

Multiplication:
☐ ☐ ☐   ☐ ☐ ☐ ☐   ☐ ☐   ☐ ☐ ☐
 × ☐   × ☐   × ☐ ☐   × ☐ ☐

Division:
☐)‾☐☐☐   ☐)‾☐☐☐☐   ☐ ☐)‾☐☐☐☐

Fractions: ☐/☐ + ☐/☐   or any other operation

Two other activities, seemingly obvious, lead to an interesting third problem.

1. Using each digit once, make the largest possible 10-digit number.
2. Using each digit once, make the largest possible sum for a 5-digit by 5-digit addition problem.
3. Using each digit once, make the smallest possible difference for a 5-digit by 5-digit subtraction problem.

The answer to (1) is 9,876,543,210. The answer to (2) is 97,531 + 86,420 = 183,951. Many pupils will think the answer to (3) is 97,531 − 86,420 = 11,111. But a much smaller difference is possible––50,123 − 49,876 = 247.

FRAMES

1. This 3-frame shows 25. What do these 3-frames show?

   |9|3|1| |9|3|1| |9|3|1| |9|3|1| |9|3|1|

   a. ____   b. ____   c. ____   d. Show 17.

2. This 5-frame shows 86. What do these 5-frames show?

   |25|5|1| |25|5|1| |25|5|1| |25|5|1| |25|5|1|

   a. ____   b. ____   c. ____   d. Show 43.

3. This 7-frame shows 74. What do these 7-frames show?

   |49|7|1| |49|7|1| |49|7|1| |49|7|1| |49|7|1|

   a. ____   b. ____   c. ____   d. Show 65.

4. This 10-frame shows 232. What do these 10-frames show?

   |100|10|1| |100|10|1| |100|10|1| |100|10|1| |100|10|1|

   a. ____   b. ____   c. ____   d. Show 432.

5. Draw a 4-frame to show 11.

6. Draw a 6-frame to show 50.

7. Draw an 8-frame to show 68.

8. If each frame above had another column on the left side, what would be the value of the column?

   a. 3-frame ____   b. 5-frame ____   c. 7-frame ____

   d. 10-frame ____   e. 4-frame ____   f. 6-frame ___   g. 8-frame ___

Frames

Mathematics teaching objectives:

. Use place value concepts to write numbers.

. Practice addition skills.

Problem-solving skills pupils might use:

. Look for patterns and/or properties.

. Make decisions based upon data.

. Create new problems by varying a given one.

Materials needed:

. None

Comments and suggestions:

. The activity is intended as a discovery lesson with several place value bases leading to the discovery that base 10 is the only one that directly translates to the standard way of writing numbers.

. Pupils may need initial help to realize that in any place value system, the value of each column is 3 times, 4 times, 5 times, etc. larger than the preceding column.

Answers:

1. a. 26   b. 23   c. 19   d.

•	• •	• •
9	3	1

2. a. 98   b. 62   c. 28   d.

•	• • •	• • •
25	5	1

3. a. 114   b. 160   c. 24   d.

•	• •	• •
49	7	1

4. a. 534   b. 225   c. 320   d.

• • • •	• • •	• •
100	10	1

5.

• • •	• • •
16	4	1

6.

•	• •	• •
36	6	1

7.

•		• • • •
64	8	1

8. a. 27   b. 125   c. 343   d. 1000

    e. 64   f. 216   g. 512

HOW MANY ARE NEEDED?

1. What is the <u>fewest</u> number of coins or bills needed to make each of the following amounts? Show each by writing the number of coins or bills needed.

	$1.00	$.50	$.25	$.10	$.05	$.01	Total Number
$ .04						4	4
$ .10							
$ .17							
$ .28							
$ .76							
$ .99							
$ 1.40							
$ 1.91							
$ 2.57							
$ 3.80							

2. What is the fewest number of coins or bills needed to make each of the following amounts? Show each by writing the number of coins or bills needed.

	$100.00	$ 10.00	$ 1.00	$.10	$.01
$ 2.31			2	3	1
$ .72					
$ 36.18					
$ 409.50					
$ 999.99					
$ 50.45					
$ 70.03					
$ 5.27					

3. Was it easier doing the top or bottom table? _____ Why? _____

How Many Are Needed?

Mathematics teaching objectives:

. Investigate the efficiency of using base 10.

. Practice addition skills.

Problem-solving skills pupils might use:

. Use a systematic list or table.

. Look for properties.

. Make explanations based on data.

Materials needed:

. None

Comments and suggestions:

. Although the activity uses decimals (in the form of money) most pupils will work this as a whole number activity in that they will think in terms of 4 cents or 4 pennies instead of .04 .

. The table at the bottom shows that base 10 notation is the only system that directly translates to the standard way of writing numbers.

Answers:

1.

	$1.00	.50	.25	.10	.05	.01	Total
$ .04						4	4
.10				1			1
.17				1	1	2	4
.28			1			3	4
.76		1	1			1	3
.99		1	1	2		4	8
1.40	1		1	1	1		4
1.91	1	1	1	1	1	1	6
2.57	2	1			1	2	6
3.80	3	1	1		1		6

2.

	$100.00	$10.00	$1.00	$ .10	$ .01
$ 2.31			2	3	1
.72				7	2
36.18		3	6	1	8
409.50	4		9	5	
999.99	9	9	9	9	9
50.45		5		4	5
70.03		7			3
5.27			5	2	7

3. The bottom table is easier because you use only $100, $10, $1, $ .10 and $ .01 .

THE SHOOTING GALLERY

1. Darlene and Sam spent much money at the fair at the shooting gallery. The two targets show how they did. Each mark is a shot. What scores did Darlene and Sam get?

Darlene

Sam

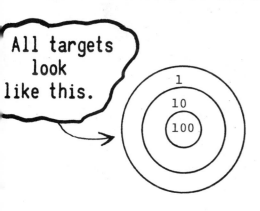 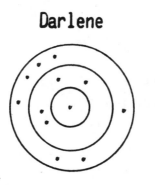 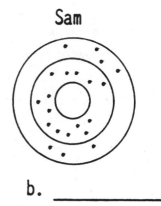

All targets look like this.

a. _____   b. _____

2. Show two ways Darlene could score 122 points.

a. 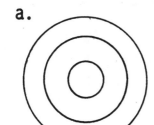   b.

3. Show how Sam could score 353 points using the fewest number of shots.

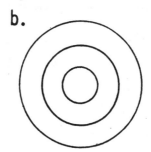

4. The booth owner said Darlene could shoot until she scored 173 points.
   a. What is the fewest number of shots she could take? _____
   b. What is the most number of shots she could take? _____

The Shooting Gallery

Mathematics teaching objectives:

. Use place value concepts to translate and write numbers.

. Investigate the efficiency of using base 10.

Problem-solving skills pupils might use:

. Use a drawing.

. Look for properties.

. Satisfy one condition at a time.

Materials needed:

. None

Comments and suggestions:

. Remind pupils that all targets on the two pages have the values of 100, 10 and 1 from inside to outside.

. Using as few shots as possible forces the answers to be written in the standard way. This illustrates the efficiency of base 10 notation.

Answers:

1. a. 147   b. 147

2. a.   b.

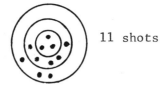

3. a.   11 shots

4. a. 11 shots   b. 173 shots assuming all shots hit the target.

The Shooting Gallery (cont.)

5. Use as few shots as possible. Show how each person could score the points listed.

    a. Albert –   342 points   e. Ellen – 3013 points
    b. Barbara –   621 points   f. Frank –   204 points
    c. Crissy – 1832 points   g. Gayle –   402 points
    d. Deron   – 1002 points

(Use small letters to show the shots.)

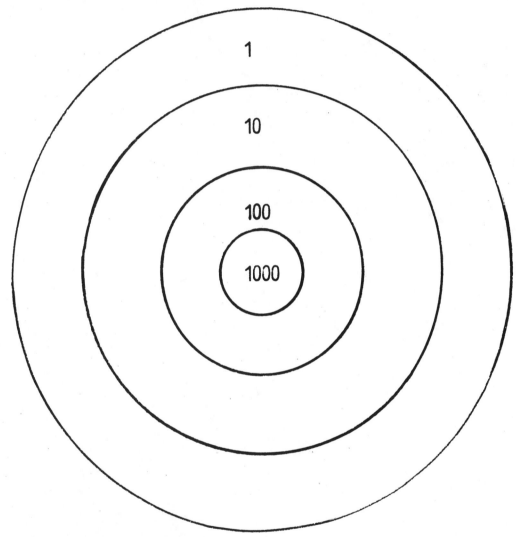

6. Who took the most shots? _____

7. Who took the fewest shots? _____

Answer:

5.

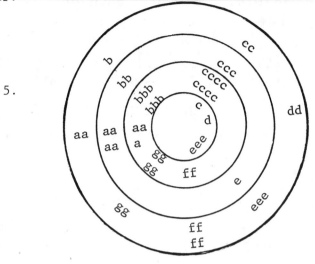

6. Crissy

7. Deron

EXCHANGES

1. Get some counters from your teacher. Use the frame at the bottom of the sheet. What is the fewest number of counters needed to show

   a. 31?   b. 113?   c. 207?

2. What number is represented on each frame below?

   (a) _____   (b) _____   (c) _____

   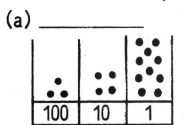   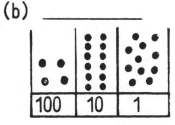   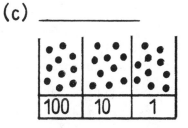

3. Use the frame and some counters. Show three different ways to represent 312. Record the number of counters you used for each one.

4. The normal representation of 423 uses nine counters.

   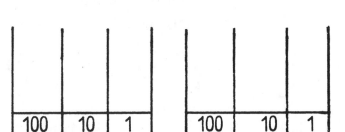

   a. Show two different ways to represent 423. Use 18 counters for each of them. Record your answers on the frames to the right.

   b. Show how to represent 423 with 27 counters.

   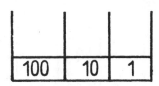

   c. What is the greatest number of counters you could use to represent 423?

<u>EXTENSION</u>   How many different ways can you represent 111 on a frame?

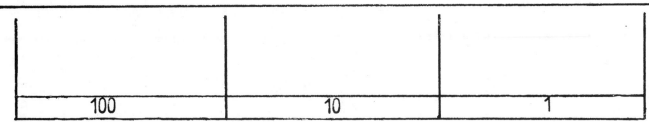

Exchanges

Mathematics teaching objectives:

. Use place value concepts to translate and write numbers.

Problem-solving skills pupils might use:

. Use a drawing.

. Make a systematic list.

Materials needed:

. A place value frame and some markers may be helpful.

Comments and suggestions:

. Pupils will need prior experiences with place value concepts
including writing numbers in standard notation from representations
shown on a place value frame. Some experiences with regrouping
by trading markers also will be helpful.

. This activity shows that a number has several meaning. For
example, 23 could mean 2 tens and 3 units or 23 units.

. Many different representations can be made for a number on a
place value frame. However, only one way - the one using the
fewest number of markers - directly translates to the standard
way of writing numbers.

Answers:

1. a. 4   b. 5   c. 9

2. a. 350   b. 531   c. 1110

3. Answers will vary.

4. a.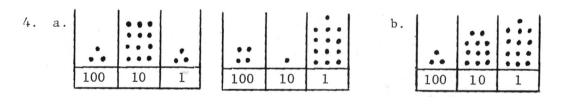

    c. 423

Extension: 13 different ways.

FIVE DIGITS

Needed: Digit cards 0-9
             2 players

1. Shuffle the cards. Deal 5 cards to each player. Write the digits each player received.

2. Player 1 ___ ___ ___ ___ ___   Player 2 ___ ___ ___ ___ ___

3. Use as many of the 5 digits, once only, as you want to get the best answer for each of these statements. Some may be impossible. When both players have written their answers, indicate the player with the best answer.

		Player 1	Player 2	Best Answer
1.	Largest possible number			
2.	Smallest possible number			
3.	Number closest to 5000			
4.	Even number closest to 200			
5.	Odd number closest to 7003			
6.	Largest number divisible by 3			
7.	Smallest number divisible by 5			
8.	Largest number divisible by 4			
9.	Smallest number divisible by 7			
10.	Smallest even number			
11.	Player 1 - Make up a statement of your own that best uses your digits.			
12.	Player 2 - Make up a statement of your own that best uses your digits.			

Five Digits

Mathematics teaching objectives:

. Practice place value skills.

. Review odd, even, largest, smallest and divisible.

Problem-solving skills pupils <u>might</u> use:

. Guess and check.

. Break a problem into parts.

. Reason from what you already know.

. Create new problems by varying an old one.

Materials needed:

. Ten digit cards marked 0-9

Comments and suggestions:

. Pupils will need to decide if a leading zero, 01478, is allowed in writing a small number. A knowledge of divisibility tests for 3, 4 and 5 would be helpful. If not, pupils might use a calculator to check out their guesses.

. An interesting problem arises if player 1 has all the even digits and player 2 has all the odd digits. Player 1 can get the best answer in only three of the ten situations.

Answers:

Answers will vary. A sample is shown below.

Player 1: <u>7</u> <u>8</u> <u>0</u> <u>1</u> <u>4</u>   Player 2: <u>9</u> <u>5</u> <u>2</u> <u>6</u> <u>3</u>

	Player 1	Player 2	Best
1.	87410	96532	2
2.	01478	23569	1
3.	4871	5236	1
4.	187	235	1
5.	7041	6953	1
6.	870	963	2
7.	10	5	2
8.	87140	96532	2
9.	7	35	1
10.	0	2	1

BIG TIC TAC TOE

Needed: Two teams
Maybe a calculator

Rules:

1. The X team picks two numbers from the bubble.

2. Multiply the numbers.

3. Find the answer below and place an X on it.

4. The O team does the same but places an O on the answer.

5. The first team to get a path connecting its two sides is the winner.

| 9 | 19 | 29 | 39 | 49 |
| 59 | 69 | 79 | 89 | |

X Team

801	3081	1131	7031	1711	931
4071	261	4361	351	5251	621
741	2581	1501	2691	551	3381
2891	2001	171	4661	2291	6141
441	5451	2301	3471	1691	1121
1311	1911	711	1421	531	3871

O Team   O Team

X Team

Special rules you have to decide on:

6. What happens if a team picks an answer that is already taken?

7. Can the path go diagonally from one square to another?

Big Tic Tac Toe

Mathematics teaching objectives:

    . Practice multiplication skills.

    . Estimate.

    . Use a calculator.

Problem-solving skills pupils might use:

    . Make reasonable estimates.

    . Record solution attempts.

    . Guess and check.

Materials needed:

    . Calculator (optional)

Comments and suggestions:

    . You may want to make a transparency and play this game as a
      whole class activity.

    . Note that pupils have to make decisions on two rules before
      play begins. If pupils decide that a turn is lost if a number
      is repeated, recording solution attempts becomes very important.

    . In this activity pupils have to make some choices. For example,
      they first estimate that 2301 is either 39 x 59 or 29 x 79.
      Then they make a guess at the correct answer.

Answers:

    . This activity has no answers other than being sure pupils have
      multiplied correctly. It is interesting to note that with the
      nine numbers given, there are exactly 36 different products
      possible.

LOWEST TO HIGHEST

1. Your teacher will read four digits to you. Write them in the four blanks of the first space below.

2. Make a two-digit by two-digit multiplication problem. Use each digit once.

3. Find the product.

4. Write the product in one of the six blanks to the right. Your goal is to finally get six products in order from lowest to highest.

5. Repeat steps 1-4 five more times.

6. To find your total score:
   a. Cross out those products that are out of order.
   b. Add the remaining products.
   c. Add a bonus of 200 if _all_ six are in order.

Lowest	a. _____
	b. _____
	c. _____
	d. _____
	e. _____
Highest	f. _____
Bonus	_____
Total	_____

Lowest To Highest

Mathematics teaching objectives:

. Practice computation skills.

. Estimate.

Problem-solving skills pupils might use:

. Make reasonable estimates.

. Guess and check.

Materials needed:

. Digit cards 0-9

. Calculator (optional)

Comments and suggestions:

. Explain to pupils the object of the activity, namely to arrange
six products from lowest to highest. But only one product will be
computed at a time. Pupils must make decisions on where to place it.
No changes can be made after a product is written in a blank.

. Pupils will need to decide whether leading zeros, like 07, will be
allowed.

. Depending on time allowed, pupils may find a product, decide it
doesn't fit, and rearrange the given four digits to get a different product which might fit.

. The element of chance in this activity allows all pupils a better
opportunity to get a high final total.

Answers:

. Answers for each pupil will vary. Perhaps they could exchange
papers and use calculators to check each other's multiplications.

LUCKY AT HIGH NOON

his retyped newspaper article appeared in the <u>Eugene Register-Guard</u>.

<u>Reedsport</u> - With a little bit of luck, Albert "Frenchie" Nimmler unseated incumbent Mayor Ron Hanson Monday.

Nimmler beat Hanson in a drawing of lots necessary to break a 774-vote tie that resulted in the Nov. 7 city election.

The drawing, held at high noon at Reedsport City Hall, was required by Oregon law as a means of breaking such a tie vote between candidates. Slips of paper numbered one through 20 were mixed in a box. Hanson and Nimmler each drew 10 slips. The man with the higher total would be the winner.

Hanson's slips totaled 95. Nimmler's totaled 115.

1. Read the article carefully.

2. Write the last names of the opponents. _____ _____

3. How many points did each have after the slips were drawn?

    _____ _____

4. Who won the election? _____

5. Do you think this is a fair way to decide the winner? _____

6. What method would you suggest? _____

    _____

7. Get a partner. Draw ten slips each. Are your results similar to the results of the election? _____

8. If you draw again, will the same person be the winner? _____ Try and find out.

9. What ten slips could Mr. Nimmler have drawn? _____

10. Is it possible, after drawing ten slips, to have a tie? _____

11. Suppose Mr. Hanson drew all the even numbers and Mr. Nimmler drew all the odd numbers. Without adding, who would have won the election? _____ By how much? _____

Lucky At High Noon

Mathematics teaching objectives:
- Practice addition and subtraction.
- Develop probability concepts.

Problem-solving skills pupils _might_ use:
- Use a model.
- Search printed matter for needed information.
- Record solution possibilities.

Materials needed:
- Twenty slips of paper numbered 1-20 for each pair of pupils.

Comments and suggestions:
- The answer to number 10 is often answered when pupils simulate the drawing in number 7. Some pair may get a tie of 105 to 105.

- Recording all the scores on the board _usually_ shows Hanson winning about the same number of times as Nimmler. When this happens most pupils accept this method as being a fair way to elect a mayor - although many feel you should flip a coin.

Answers:

2. Nimmler and Hanson

3. Nimmler 115, Hanson 95

4. Nimmler

5. and 6. Answers will vary.

7. Either yes or no

8. Maybe

9. There are many possibilities - one is 20,19,18,17,11,10,8,6,4,2 .

10. Yes, $20 + 19 + 18 + 17 + 16 + 5 + 4 + 3 + 2 + 1 = 105$

and

$15 + 14 + 13 + 12 + 11 + 10 + 9 + 8 + 7 + 6 = 105$

11. Hanson would win by 10 .

HELPING OUT
(A Teacher Idea)

1. Prepare five squares as shown below. Number as indicated. Make the squares 15 cm by 15 cm.

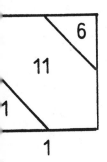 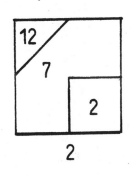 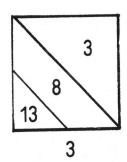 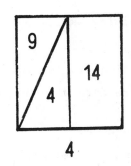 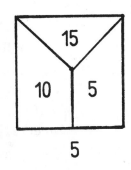

2. Cut out the pieces and place in five envelopes. Pieces 1, 2, 3 go together. Also 4, 5, 6 and 7, 8, 9 and 10, 11, 12 and 13, 14, 15.

3. The task of five pupils is to reassemble the five squares following these rules.

    a. No talking is allowed.
    b. A pupil may <u>give</u> a piece to another pupil.
       Pieces may not be <u>taken</u> from others.

Note: Square 5 is the most difficult. Some hints may be needed.

4. After all squares are completed, have the team of five assign a fraction value to each piece. An assembled square is the unit.

Helping Out

Mathematics teaching objectives:

. Recognize shapes needed to make a square.

. Name fractional parts of a square.

Problem-solving skills pupils might use:

. Use a model.

. Break a problem into parts.

. Share data with other persons.

Materials needed:

. Shapes as indicated by the drawings on the pupil page.

Comments and suggestions:

. Square 5 is the most difficult to complete. Pupils often place pieces 5 and 10 with the long sides together.

If the activity bogs down, you might place the pieces together in the correct way. Or you might allow pupils to ask for a particular piece by number.

. The fraction name for each piece can be determined by actual covering of pieces by known pieces. For example, piece 2 is easily determined to be $\frac{1}{4}$ of a square. Pieces 1, 6, 12, and 13 are each $\frac{1}{8}$ of a square because two of them will exactly cover piece 2. Then piece 7 can be exactly covered by pieces 2, 1, 6, and 12, making it $\frac{5}{8}$ of a square.

. Another way of doing piece 7 is to subtract $\frac{1}{4}$ and $\frac{1}{8}$ from 1 to get $\frac{5}{8}$.

Answers:

1. $\frac{1}{8}$ 6. $\frac{1}{8}$ 11. $\frac{3}{4}$

2. $\frac{1}{4}$ 7. $\frac{5}{8}$ 12. $\frac{1}{8}$

3. $\frac{1}{2}$ 8. $\frac{3}{8}$ 13. $\frac{1}{8}$

4. $\frac{1}{4}$ 9. $\frac{1}{4}$ 14. $\frac{1}{2}$

5. $\frac{3}{8}$ 10. $\frac{3}{8}$ 15. $\frac{1}{4}$

NUMBERS IN BOXES

The numbers inside the dotted boxes can be found by looking for
patterns. Discover the pattern and then finish problems c and d.

1.

a.

3	2	
1	1	5
2	3	6

b.

8	9	
2	3	17
3	4	12

c.

1	1	
5	6	

d.

2	3	
5	4	

2.

a.

3	10	
1	5	13
4	6	12

b.

15	4	
5	1	19
8	6	24

c.

4	3	
6	4	

d.

1	5	
2	6	

3.

a.

4	1	
2	1	5
3	6	6

b.

15	4	
5	2	19
6	9	18

c.

4	1	
9	3	9

d.

1	1	
3	3	

4.

Create your own numbers in boxes. Do a problem like 1, like 2,
and like 3.

a.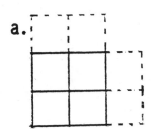

b.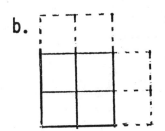

c.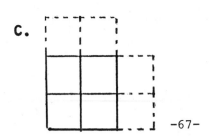

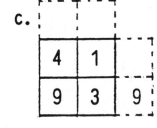

<u>Numbers</u> <u>In</u> <u>Boxes</u>

Mathematics teaching objectives:

. Practice addition and multiplication.

Problem-solving skills pupils <u>might</u> use:

. Look for patterns.

. Create other problems which can be solved by a certain procedure.

Materials needed:

. None.

Comments and suggestions:

. Pupils may recognize many different patterns. As long as the pupil can explain the pattern (and it is valid), accept it as correct.

. Few, if any, pupils (or your colleagues) will recognize this as the addition algorithm for fractions. For example 3-a could be rewritten as

$$\frac{2}{3} = \frac{4}{6}$$
$$+ \ \frac{1}{6} = \frac{1}{6}$$
$$\frac{5}{6}$$

. Problem 1 has no common factor in the denominator. Problem 2 has a common factor of 2. Problem 3 has a common factor of 3.

Answers:

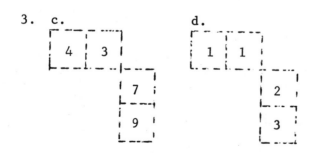

1. c.
```
  6  | 5 |
         | 11 |
         | 30 |
```
   d.
```
  8  | 15 |
          | 23 |
          | 20 |
```
2. c.
```
  8  | 9 |
        | 17 |
        | 12 |
```
   d.
```
  3  | 5 |
```

3. c.
```
  4 | 3 |
       | 7 |
       | 9 |
```
   d.
```
  1 | 1 |
       | 2 |
       | 3 |
```
4. Answers will vary.

FRACTION MAGIC SQUARES

Finish these magic squares. Each row, column, and diagonal adds to the same magic sum.

1.

	$\frac{2}{15}$	
	$\frac{10}{15}$	
	$\frac{18}{15}$	$\frac{4}{15}$

Magic Sum _____

2.

		$\frac{2}{10}$
$\frac{8}{10}$	$\frac{3}{10}$	$\frac{4}{10}$

Magic Sum _____

3.

	$\frac{11}{12}$	$\frac{1}{2}$
	$\frac{7}{12}$	
$\frac{2}{3}$		

Magic Sum _____

4.

$\frac{1}{16}$			$\frac{1}{4}$
	$\frac{3}{8}$		
	$\frac{5}{8}$	$\frac{11}{16}$	$\frac{5}{16}$
$\frac{13}{16}$	$\frac{3}{16}$		1

Magic Sum _____

5.

$\frac{17}{25}$	$\frac{24}{25}$		$\frac{8}{25}$	
$\frac{23}{25}$	$\frac{5}{25}$			$\frac{16}{25}$
	$\frac{6}{25}$	$\frac{13}{25}$		$\frac{22}{25}$
$\frac{10}{25}$			$\frac{21}{25}$	$\frac{3}{25}$
$\frac{11}{25}$	$\frac{18}{25}$			$\frac{9}{25}$

Magic Sum _____

6. Use these fractions to create a magic square with a sum of $2\frac{1}{2}$.

$\frac{1}{6}$, $\frac{1}{3}$, $\frac{1}{2}$, $\frac{2}{3}$, $\frac{5}{6}$, 1, $1\frac{1}{6}$, $1\frac{1}{3}$

	$1\frac{1}{2}$	

7. Create your own magic square.

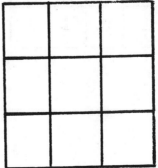

Fraction Magic Squares

Mathematics teaching objectives:

. Practice addition and subtraction of fractions.

Problem-solving skills pupils might use:

. Break a problem into parts.
. Look for patterns.

Materials needed:

. None

Comments and suggestions:

. Pupils need to examine each magic square to find a row, column or diagonal that is completed in order to find the magic sum. After that other spaces may be filled in by adding and subtracting.

. Some pupils will find squares 3, 4 and 6 easier if fractions are re-written with a common denominator. In fact, with the denominators the same, the magic squares can be done like whole number magic square by just looking at the numerators of the fractions.

Answers:

1.

$\frac{16}{15}$	$\frac{2}{15}$	$\frac{12}{15}$
$\frac{6}{15}$	$\frac{10}{15}$	$\frac{14}{15}$
$\frac{8}{15}$	$\frac{18}{15}$	$\frac{4}{15}$

$\frac{30}{15}$ or 2

2.

$\frac{6}{10}$	$\frac{7}{10}$	$\frac{2}{10}$
$\frac{1}{10}$	$\frac{5}{10}$	$\frac{9}{10}$
$\frac{8}{10}$	$\frac{3}{10}$	$\frac{4}{10}$

$\frac{15}{10}$ or $1\frac{1}{2}$

3.

$\frac{1}{3}$	$\frac{11}{12}$	$\frac{1}{2}$
$\frac{3}{4}$	$\frac{7}{12}$	$\frac{5}{12}$
$\frac{2}{3}$	$\frac{1}{4}$	$\frac{5}{6}$

$\frac{21}{12}$ or $1\frac{3}{4}$

4.

$\frac{1}{16}$	$\frac{15}{16}$	$\frac{7}{8}$	$\frac{1}{4}$
$\frac{3}{4}$	$\frac{3}{8}$	$\frac{7}{16}$	$\frac{9}{16}$
$\frac{1}{2}$	$\frac{5}{8}$	$\frac{11}{16}$	$\frac{5}{16}$
$\frac{13}{16}$	$\frac{3}{16}$	$\frac{1}{8}$	1

$\frac{34}{16}$ or $2\frac{1}{8}$

5.

$\frac{17}{25}$	$\frac{24}{25}$	$\frac{1}{25}$	$\frac{8}{25}$	$\frac{15}{25}$
$\frac{23}{25}$	$\frac{5}{25}$	$\frac{7}{25}$	$\frac{14}{25}$	$\frac{16}{25}$
$\frac{4}{25}$	$\frac{6}{25}$	$\frac{13}{25}$	$\frac{20}{25}$	$\frac{22}{25}$
$\frac{10}{25}$	$\frac{12}{25}$	$\frac{19}{25}$	$\frac{21}{25}$	$\frac{3}{25}$
$\frac{11}{25}$	$\frac{18}{25}$	$\frac{25}{25}$	$\frac{2}{25}$	$\frac{9}{25}$

$\frac{65}{25}$ or $2\frac{15}{25}$ or $2\frac{3}{5}$

6.

$1\frac{1}{3}$	$\frac{1}{6}$	1
$\frac{1}{2}$	$\frac{5}{6}$	$1\frac{1}{6}$
$\frac{2}{3}$	$1\frac{1}{2}$	$\frac{1}{3}$

R¹D¹C⁰U^LOUS RU^LER

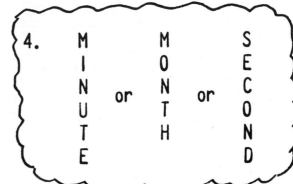

E F P O C T X Q H I J B W L S K A N Y V R U M D Z G

According to the ruler, <u>PAM</u> is $\frac{3}{4} + 4\frac{1}{4} + 5\frac{3}{4} = 10\frac{3}{4}$

According to this ruler, which is "longer?"   How much "longer?"

1. C   D
    A   or   O
    T   G

2. W   P
    O   L
    R   or   A
    K   Y

According to the ruler, which is "longest?"

3. I   F   M
    N   O   I
    C or O or L
    H   O   L
          T   E

4. M   M   S
    I   O   E
    N   N   C
    U or   T or   O
    T   H   N
    E   D

5. Who has the "longest" name in the class?

6. What is the "longest" 3-letter word on the ruler?

7. Create your own question to ask.

Ridiculous Ruler

Mathematics teaching objectives:

    . Practice addition of fractions, both mixed and proper.

    . Read a number line (ruler) marked in fourths.

Problem-solving skills pupils might use:

    . Use a drawing.

    . Break a problem into parts.

    . Create new problems by varying a given one.

Materials needed:

    . None

Comments and suggestions:

    . Introduce the activity with the overhead projector. Use the
      letters shown or let the pupils arrange the letters as they want.

    . The words on the activity are written vertically. This helps
      pupils to place the values vertically and makes adding easier.

Answers:

    1. Cat - 7, Dog - 13 $\frac{1}{2}$. Dog is more by 6 $\frac{1}{2}$.

    2. Work - 13 $\frac{1}{2}$, Play - 13 $\frac{1}{4}$. Work is more by $\frac{1}{4}$.

    3. Inch - 10 $\frac{1}{2}$, Foot - 4, Mile - 12. Mile is longest.

    4. Minute - 20, Month - 15, Second - 16 $\frac{3}{4}$. Minute is longest.

    5. Answers will vary.

    6. Dug is the longest 3-letter word on this ruler:

$$6 + 5 \frac{1}{2} + 6 \frac{1}{2} = 18$$

FRACTION PATTERNS

1.

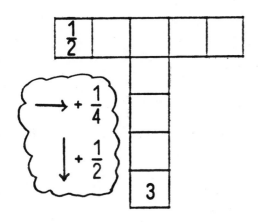

2.

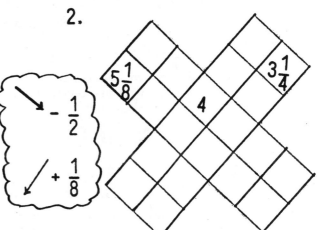

3.

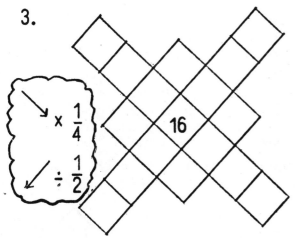

4.

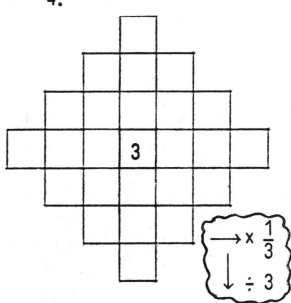

5. Create a fraction pattern of your own.

Fraction Patterns

Mathematics teaching objectives:

. Practice basic skills with fractions.

Problem-solving skills pupils <u>might</u> use:

. Look for patterns.
. Satisfy one condition at a time.
. Work backwards.

Materials needed:

. None

Comments and suggestions:

. Start a problem with your pupils so they understand that one operation is being done in one direction and another in the other direction.

. Beware of the " $\div \frac{1}{2}$." Many pupils will interpret this as " $\div 2$."

The problem may be an opportunity to talk about the following:

When multiplying by a number

greater than 1, the product gets larger.
equal to 1, the product stays the same.
less than 1, the product gets smaller.

When dividing by a number

greater than 1, the quotient gets smaller.
equal to 1 , the quotient stays the same.
less than 1, the quotient gets larger.

. If pupils create their own patterns, they may use only + and − or x and ÷ in a problem.

Answers:

1.

2.

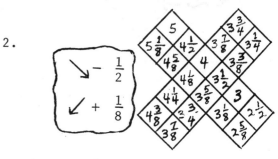

3.

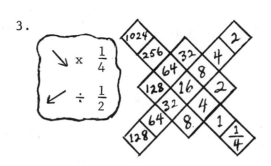

4.

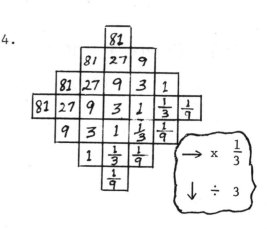

DECIMAL MAZES

Find a path through each maze that gives the sum at the end. Some mazes have more than one correct path.

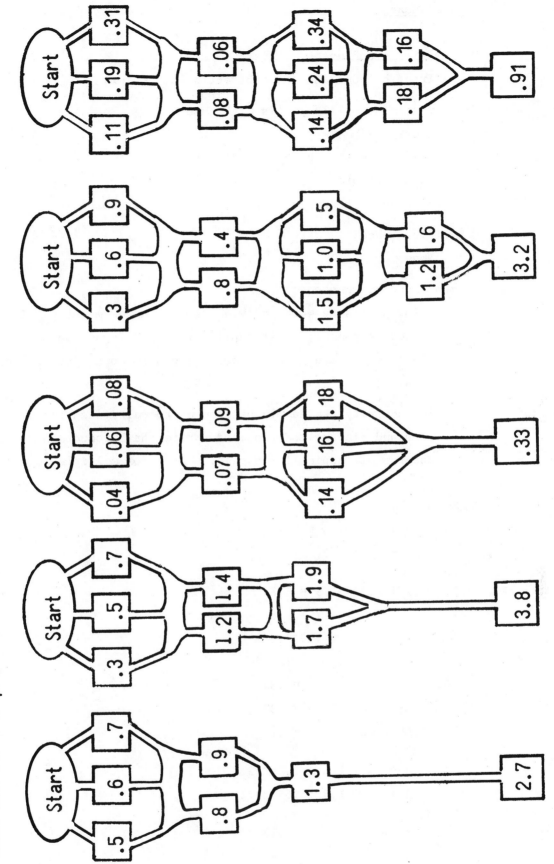

PSM 81

Decimal Mazes

Mathematics teaching objectives:

. Practice addition and subtraction of decimals.

Problem-solving skills pupils might use:

. Guess and check.

. Solve an easier but related problem.

. Work backwards.

. Find another answer when more than one is possible.

Materials needed:

. None

Comments and suggestions:

. Allow pupils to work individually for a few minutes. Then call for correct reponses. Most pupils will have found one correct answer. When several are given, encourage others to try to find as many as possible.

. The activity can easily be changed to multiplication by choosing a path and finding the product. Write that product in the last box and run off copies for pupils. Watch for multiple solutions!

Answers:

1. $.5 + .9 + 1.3 = 2.7$
    $.6 + .8 + 1.3 = 2.7$

2. $.7 + 1.2 + 1.9 = 3.8$
    $.7 + 1.4 + 1.7 = 3.8$
    $.5 + 1.4 + 1.9 = 3.8$

3. $.06 + .09 + .18 = .33$
    $.08 + .07 + .18 = .33$
    $.08 + .09 + .16 = .33$

4. $.6 + .4 + 1.0 + 1.2 = 3.2$
    $.3 + .8 + 1.5 + .6 = 3.2$

5. $.31 + .08 + .34 + .18 = .91$

DECIMAL MAGIC SQUARES

Finish these magic squares. Each row, column, and diagonal adds to the same magic sum.

1.

8.5	1.5	
	5.5	
		2.5

Magic Sum = _____

2.

5.2		
5.6		4
3.6		

Magic Sum = _____

3.

		.93
.65	1.35	.37

Magic Sum = _____

4.

3.15		1.47
2.73		1.05

Magic Sum = __6.3__

5.

1.3			2.2
	2.8		
	4	4.3	2.5
4.9	1.9		5.8

Magic Sum = _____

6.

.55	.5		1.15	.85
.9		.3	.25	1.2
		.65		
	1.05			.4
.45	.15	1.1	.8	

7. Create your own decimal magic square.

Decimal Magic Squares

Mathematics teaching objectives:

. Practice addition and subtraction of decimals.

Problem-solving skills pupils __might__ use:

. Break a problem into parts.

. Look for patterns.

Materials needed:

. None

Comments and suggestions:

. Pupils need to examine each magic square to find a row, column or diagonal that is completed in order to find the magic sum. After that, other spaces may be filled in adding and subtracting.

. To do (4) pupils need to know (perhaps from previous work with magic squares) that the middle number can be found by finding the average of the four corner squares. Or they can work backward from the magic sum.

Answers:

1.

8.5	1.5	6.5
3.5	5.5	7.5
4.5	9.5	2.5

__16.5__

2.

5.2	3.2	6.0
5.6	4.8	4
3.6	6.4	4.4

__14.4__

3.

1.21	.23	.93
.51	.79	1.07
.65	1.35	.37

__2.37__

4.

3.15	1.68	1.47
.42	2.1	3.78
2.73	2.52	1.05

__6.3__

5.

1.3	5.5	5.2	2.2
4.6	2.8	3.1	3.7
3.4	4	4.3	2.5
4.9	1.9	1.6	5.8

__14.2__

6.

.55	.5	.2	1.15	.85
.9	.6	.3	.25	1.2
1.25	.95	.65	.35	.05
.1	1.05	1	.7	.4
.45	.15	1.1	.8	.75

__3.25__

DECIMAL TIC TAC TOE

1. This game needs two players.
2. Pick a number from column 1 and a number from column 2.
3. Multiply the numbers. Find the answer and mark it with an
   X or O.
4. Play each game like regular Tic Tac Toe and try to get three in a row.

	1	2
	1.1	4.1
	2.1	5.1
	3.1	6.1

18.91	6.71	15.81
5.61	12.71	10.71
12.81	8.61	4.51

	1	2
	10	4.5
	5.8	7.1
	3.2	9.9

31.68	45	26.1
99	14.4	57.42
41.18	71	22.72

	1	2
	4.2	6
	9.3	5.5
	.5	.8

2.75	3.36	7.44
55.8	.4	23.1
3	51.15	25.2

	1	2
	4.2	6.4
	3.9	5.8
	8.1	2.9

12.18	24.96	23.49
46.98	26.88	11.31
24.36	22.62	51.84

Create your own decimal
Tic Tac Toe.

1	2

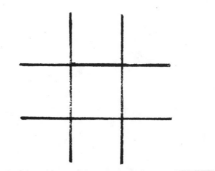

Decimal Tic Tac Toe

Mathematics teaching objectives:
- . Practice multiplication of decimals.
- . Use a calculator (optional).

Problem-solving skills pupils <u>might</u> use:
- . Make reasonable estimates.
- . Record solution attempts.
- . Guess and check.

Materials needed:
- . Calculator (optional)

Comments and suggestions:
- . Recording the numbers seclected from columns 1 and 2 is important to avoid choosing those same numbers later in the game. The recordings may also help in making reasonable guesses later in the game.

Answers:

- . This activity has no answers. Pupils should be sure they have multiplied correctly. It is interesting to note that with the six numbers given, exactly nine different products are possible.

TEST ANSWERS

Harriet's test answers are shown below. Estimate only to decide which answers are not correct. Mark an X beside the wrong answers.

1. $39.82 + 5.76 =$ _45.58_
2. $2.38 + 6.5 + 9.27 =$ _12.30_
3. $3.024 + 5.718 + 11.135 =$ _19.877_
4. $6.23 + 16.84 + 19.21 =$ _42.28_
5. $17.91 + 30.2 + 23.62 =$ _44.55_
6. $14.98 - 2.76 =$ _12.22_
7. $8.392 - 5.14 =$ _3.252_
8. $32.23 - 29.96 =$ _17.73_
9. $62.824 - 39.98 =$ _37.164_
10. $39.84 - 26.83 =$ _13.01_
11. $.3 \times 932 =$ _279.6_
12. $6 \times 25.6 =$ _153.6_
13. $3.9 \times 14 =$ _54.6_
14. $2.3 \times 5.2 =$ _119.6_
15. $4.32 \times 8.37 =$ _3615.84_
16. $8.324 \div 4 =$ _2.081_
17. $92.8 \div 8 =$ _11.6_
18. $35.25 \div 15 =$ _2.35_
19. $18.24 \div 3.2 =$ _.57_
20. $480.33 \div 59.3 =$ _.81_

---

1. Work the ones Harriet missed to find the correct answers.

2. Explain why she missed the ones she did.

Test Answers

Mathematics teaching objectives:
. Practice basic skills with decimals.
. Estimate.

Problem-solving skills pupils might use:
. Make reasonable estimates.
. Detect and correct errors.
. Make explanations based upon data.

Materials needed:
. None

Comments and suggestions:
. The activity is designed to really force pupils to estimate.
Encourage no use of computation skills except those needed to
estimate.
. Harriet missed two problems from addition, two from subtraction,
two from multiplication and two from division. For each operation
the mistake she made was the same. To determine the error, pupils
should write the problem in the standard way, e.g. write the
addition problems in vertical form.

Answers:

Harriet missed problems 2, 5, 8, 9, 14, 15, 19, 20.

1. (2) 18.15   (5) 71.73   (8) 2.27   (9) 22.844

    (14) 11.96   (15) 36.1584   (19) 5.7   (20) 8.1

2. The mistake in addition was not lining up the decimal points.
    The mistake in subtraction was subtracting the smaller number
    from the larger number without doing any regrouping.
    The mistake in multiplication was bringing the decimal point
    straight down.
    The mistake in division was moving the decimal point straight
    up.

Grade 6

III. STORY PROBLEMS

III. STORY PROBLEMS

Story problems are an
important part of problem
solving in mathematics.
They are used throughout
mathematics and science
courses as applications
for concepts and skills.
Their purpose is to help
pupils transfer what they
are learning in the classroom to everyday life and
later to careers.

The teaching of story problems has been criticized in several ways.
One complaint is that teaching story problems often is the only attempt to .
teach problem solving. As this book shows, problem solving
is much more than just story problems. Another criticism is that a page of
story problems may use only one skill, such as subtraction. In this case,
pupils can decide what to do to solve problem 1 and avoid reading or understanding the rest. A third criticism is that pupils are not helped to
understand the story problems but are unintentionally encouraged to race to
the answers. These things can lead pupils to develop their own incorrect rules.
As a spoof, Joe Dodson, Mathematics Supervisor for the Winston-Salem/Forsyth
County Schools, illustrates in the North Carolina State Math Newsletter:

"A STUDENT'S GUIDE TO PROBLEM SOLVING

"Rule 1: If at all possible, avoid reading the problem. Reading
the problem only consumes time and causes confusion.

"Rule 2: Extract the numbers from the problem in the order in
which they appear. Be on the watch for numbers written in words.

"Rule 3: If rule 2 yields three or more numbers, the best bet
for getting the answer is adding them together.

"Rule 4: If there are only two numbers which are approximately
the same size, then subtraction should give the best results.

"Rule 5: If there are only two numbers in the problem and one is
much smaller than the other, then divide if it goes evenly--
otherwise, multiply.

"Rule 6: If the problem seems like it calls for a formula, pick a formula that has enough letters to use all the numbers given in the problem.

"Rule 7: If the rules 1-6 don't seem to work, make one last desperate attempt. Take the set of numbers found by rule 2 and perform about two pages of random operations using these numbers. You should circle about five or six answers on each page just in case one of them happens to be the answer. You might get some partial credit for trying hard.

"Rule 8: Never, never spend too much time solving problems. This set of rules will get you through even the longest assignments in no more than ten minutes with very little thinking."

The criticisms listed previously can be avoided by teaching problem solving in many ways, by providing sets of problems requiring several different skills and by giving activities where the objective is to understand the story problems, not "race to the answer." These ideas are incorporated in this section.

The five activities in this section primarily use whole numbers and money. All four operations are involved. Although estimation is a main point of several activities, pupils should acquire the habit of checking the reasonableness of each answer. Three activities include ideas not often found in texts:

. pupils supply missing information needed to solve a problem.

. pupils identify extra information in a problem.

. pupils create problems to match given solutions.

SPORTING GOODS

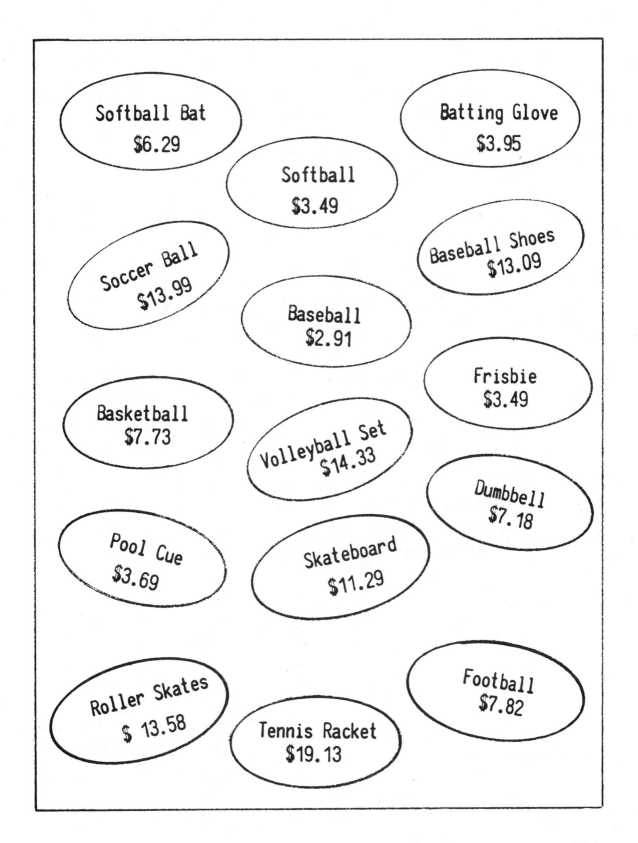

Softball Bat $6.29

Batting Glove $3.95

Softball $3.49

Soccer Ball $13.99

Baseball Shoes $13.09

Baseball $2.91

Frisbie $3.49

Basketball $7.73

Volleyball Set $14.33

Dumbbell $7.18

Pool Cue $3.69

Skateboard $11.29

Roller Skates $ 13.58

Football $7.82

Tennis Racket $19.13

PSM 81

Sporting Goods

Mathematics teaching objectives:

  . Estimate.

  . Do mental computations

  . Compute with money amounts.

Problem-solving skills pupils might use:

  . Make reasonable estimates.

Materials needed:

  . Calculator (optional)

Comments and suggestions:

  . An overhead transparency of the prices would eliminate the need for each pupil to have a copy of the first page. Perhaps a bulletin board display of the items would be helpful.

  . This activity emphasizes how important it is to undertand the problem. No correct, exact answer is wanted - only an estimate to determine if each person has enough money to make the purchase. In fact, if pupils are required to find an exact answer, the need for estimation is removed.

  . A discussion about how pupils feel about their extimates is important. Emphasize that some estimates are closer than others but seldom is any estimate wrong - unless it is extremely far off.

Answers:

1.	Yes		7.	No
2.	No		8.	No
3.	Yes		9.	Yes
4.	No		10.	No
5.	Yes		11.	Yes
6.	Yes		12.	After the $13.58 purchase

SPORTING GOODS

Ten friends are going to the Sporting Goods Store. The amount each has to spend is shown. The items each wants to buy are listed. Estimate to determine whether each has enough money.

		Money	Items	Enough Money?
1.	Beth	$ 25	Soccer Ball and Basketball	
2.	Betsy	$ 14	Skateboard and Frisbie	
3.	Ben	$ 15	Dumbbell and Football	
4.	Bill	$ 25	Basketball and Tennis Racket	
5.	Becky	$ 40	Volleyball Set, Tennis Racket and Frisbie	
6.	Barb	$ 30	Softball, Softball Bat, Batting Glove, and Baseball Shoes	
7.	Bev	$ 20	Two Pool Cues and Roller Skates	
8.	Bob	$ 20	Baseball, Baseball Shoes and two Batting Gloves	
9.	Boomer	$ 40	Roller Skates, Skateboard, and two Dumbbells	
10.	Bert	$ 40	Softball, Volleyball Set, Baseball, Soccer Ball, Football, & Basketball	

11. If all the friends put their money together, could they buy two of each item? _____

12. Bonnie had $100 to spend. These are the prices. She should have estimated the cost of each item as she bought it. Estimate to determine when Bonnie should have stopped.

$13.99, $7.73 $11.29 $7.18 $13.09 $3.69 $7.82, $6.29
$3.95, $13.58, $19.13, $3.49, $2.91

HSM 81

OPERATION, PLEASE

For each problem, circle the operation needed to find the answer.
Some questions may have more than one correct answer.

a. The hair on your head grows about 15 mm per month. How much
   will your hair grow in 12 months?

        Add   Subtract   Multiply   Divide

b. Glenda has $78.47 in her savings account. She uses $18.79 to
   buy a pair of running shoes. How much is in her account now?

        Add   Subtract   Multiply   Divide

c. The cafeteria has 17 tables that can seat 204 students.
   How many students can sit at each table?

        Add   Subtract   Multiply   Divide

d. Before the trip, the odometer read 15,419 miles. The trip
   covered a distance of 1,522 miles. What does the odometer
   read now?

        Add   Subtract   Multiply   Divide

e. Tony picked 138 apples. He gave 23 of them to his neighbor,
   Mr. Dutch, for a pie. How many apples does Tony have now?

        Add   Subtract   Multiply   Divide

f. The school store sells 48 packages of raisins each school day.
   How many packages are sold in a month containing 21 school days?

        Add   Subtract   Multiply   Divide

g. Meredith types 80 words per minute. How long would it take
   her to type a 2480 word essay?

        Add   Subtract   Multiply   Divide

Operation, Please

Mathematics teaching objective:

. Choose the correct operation(s) to solve a problem.

Problem-solving skills pupils might use:

. Make decisions based upon data.

. Find another answer when more than one is possible.

Materials needed:

. None

Comments and suggestions:

. Asking pupils to just decide on the operation to do is a way of emphasizing the understanding of the problem. No answers are needed. If you want to have pupils go ahead and find the answers, partial credit should be given for the appropriate choice of operation.

. Some problems, especially those needing two operations, can be done in more than one way. If pupils can reasonably defend their choice of operation, credit should be given.

Answers:

a.	Multiply	g.	Divide
b.	Subtract	h.	Subtract
c.	Divide	i.	Add or Multiply, then Add
d.	Add	j.	Subtract or Add, then Subtract
e.	Subtract	k.	Multiply, then Divide
f.	Multiply	l.	Add, then Divide

Operation, Please (cont.)

h. In the magazine drive Sharon sold 83 subscriptions. Bruce
    sold 57 subscriptions. How many more did Sharon sell?

         Add   Subtract   Multiply   Divide

For each problem, circle the operation to do first. Then circle
the operation to do second.

i. Wanda bought two racquetballs at $1.10 each and a racquet
    for $26.95. How much did she spend?

         First: Add   Subtract   Multiply   Divide
         Second: Add   Subtract   Multiply   Divide

j. Juan has $20.00 to shop for presents. He plans to buy his
    sister a blouse for $8.98. Then he wants to buy his dad a
    special fishing lure for $7.83. How much will Juan have left?

         First: Add   Subtract   Multiply   Divide
         Second: Add   Subtract   Multiply   Divide

k. Tammy's softball team bought six softballs at $3.36 each.
    How much should each of the 12 team members pay?

         First: Add   Subtract   Multiply   Divide
         Second: Add   Subtract   Multiply   Divide

ℓ. Jim's scores on five math quizzes were 78, 85, 62, 94, and 96.
    What was his average score?

         First: Add   Subtract   Multiply   Divide
         Second: Add   Subtract   Multiply   Divide

MORE INFORMATION, PLEASE

1. Mary bought 3 record albums costing $5.98 each. How much change did she receive?

    a. Can you solve this problem? _____

    b. What information is missing? _____

    c. Make up your own information. Be reasonable. _____

    d. Solve the problem. Mary's change is _____.

2. For each problem, write what information is missing, if any. Then create your own reasonable information and solve the problem.

    a. Ms. Perez' class has 13 boys. Two girls moved away. How many pupils are in the class now?

    b. Susan's family each rode their bicycles a total of 15 kilometres. Altogether how many kilometres did they ride?

    c. Mr. Simpson decided to have meatloaf for supper. He bought $2\frac{1}{2}$ pounds of hamburger costing $1.50 per pound. How much did the meat cost?

    d. John's two notebooks and five pencils cost $1.60. How much did each notebook cost?

    e. Lonnie's doctor told her to take 3 pills a day until the pills were gone. If the pills cost $4.80, how many days will they last?

    f. If gasoline costs 30.8 cents per litre, how much will it cost Connie to fill the tank of her pickup?

    g. Eddy babysat for five days in a row. Mr. Jones paid him $25. How much did he make on Monday?

    h. Bananas were on sale for 35¢ a pound. Mrs. Phie bought a bunch of 12 bananas. How much did they cost?

-95-

<u>More</u> <u>Information</u>, <u>Please</u>

Mathematics teaching objectives:

. Solve single step word problems.
. Compute with money amounts.
. Practice basic fact operations.

Problem-solving skills pupils <u>might</u> use:

. Create data needed to solve a problem.
. Use math symbols to describe situations.

Materials needed:

. None

Comments and suggestions:

. This page prevents pupils from racing to an answer. They must first decide what is missing. Here is a chance to emphasize how important it is to understand the problem. Pupils can be given credit for identifying what's missing <u>before</u> they are asked to find the answers. In fact, you might not have them find the answers at all for this activity.

. Using the overhead projector or a sheet for each pupil, allow them to ponder and struggle with the first problem. Discuss the needed information and then solve the problem.

. Encourage pupils to use reasonable information while solving the remainder of the activity. You may want pupils to write a sentence describing the needed information.

. Many pupils will think that problem <u>h</u> can be solved as is!

. Pupils could switch papers to check computation by hand or calculator.

Answers:

1. a. Not now.   b. How much did Mary give the salesclerk?
    c. and d. Answers will vary.

2. a. How many girls were in the class?
    b. How many people are in Susan's family?
    c. Nothing is missing. The answer is $3.75 .
    d. How much does a pencil cost?
    e. How many pills did Lonnie get?
    f.. How many litres are needed to fill the gas tank?
    g. Did Eddy receive the same amount for each day?
    h. How many pounds did the 12 bananas weigh?

TOO MUCH!

For each problem, cross out the unneeded information. Then solve the problem.

1. Mary's dog, Phidough, weighs 5 kilograms and is 82 centimetres long. Arthur, her cat, is only 46 centimetres long. How much longer is Phidough?

2. Each Saturday the Jiffy Bread Company has a sale on left-over bread. White bread sells for 48¢ a loaf. Wheat bread sells for 56¢ a loaf. How much does Mr. Carlos spend if he buys 12 loaves of wheat bread?

3. Last year Lisa averaged 12.3 points a game and made 47% of her shots. This year she averaged 17.1 points while making 51% of her shots. How much more is her average this year?

4. The Corner Fruit Stand sells apples at 43¢ a pound, bananas at 23¢ a pound, and oranges at 38¢ a pound. How much would 3 pounds of apples, 12 bananas weighing 3 pounds, and 6 pounds of oranges cost?

5. Rose's time for a 26-mile, 385-yard marathon was 3 hours 39 minutes. Dennis' time was 3 hours 53 minutes. What was the difference between their times?

6. A ream of paper contains 500 sheets. Paper comes 10 reams to a box. Pleasant Mountain School ordered 417 boxes. How many reams of paper did the school buy?

Too Much!

Mathematics teaching objectives:

- Solve word problems.
- Compute with money amounts.
- Practice basic fact computations.

Problem-solving skills pupils might use:

- Eliminate extraneous information.
- Use math symbols to describe situations.

Materials needed:

- None

Comments and suggestions:

- Most problems outside the mathematics text contain more information than is needed. This activity focuses on this attribute of problems. Pupils will need to read each problem carefully, locate the unnecessary information and cross it out. To focus pupils' attention on the importance of understanding the problem, credit can be given for crossing out the extra information before pupils finish solving the problem.
- A collection of pupil created problems could be dittoed and used as a follow-up to this activity.

Answers:

	Unneeded Information	Answer
1.	Phidough weighs 5 kilograms.	36 centimetres
2.	White bread sells for 48¢ a loaf.	$6.72
3.	47% and 51% of her shots	4.8 points
4.	12	$4.26
5.	26 mile, 385 yard	14 minutes
6.	A ream of paper contains 500 sheets.	4170 reams
7.	3 bars of soap at 52¢ each	$6.96
8.	Each person pays $3.50 to attend the movie.	20 people
9.	2-inch wide	2 yards
10.	The barn was 20 m long x 7 m high x 30 m wide.	1232 gallons
11.	He didn't get 3 pages done.	133 pages

7. Earl's mother sent him to the store to buy 2 gallons of milk at $1.58 a gallon, 3 loaves of bread at 55¢ each, 5 pounds of apples at 43¢ a pound, and 3 bars of soap at 52¢ each. How much did the food items cost?

8. Cinema III can seat 500 people in 25 rows of seats. Each person pays $3.50 to attend the movie. How many people can be seated in each row?

9. Tammy made 2-inch wide sweat bands for the 12 members of the girl's basketball team. Each band used $\frac{1}{6}$ of a yard of terrycloth. How much material was needed in all?

10. Farmer Brown wanted to paint his blue barn white. The barn is 20 m long x 7 m high x 30 m wide. He knew he was a sloppy painter and would spill 2 litres for every <u>one</u> he got on the barn. So he bought 3,696 litres of paint. How much paint did it take to cover the barn?

11. Bobby Booker loved homework so he begged his teachers for extra work to do. He had 7 classes and <u>each</u> teacher reluctantly gave him the following assignments: 4 pages on Monday, 6 on Tuesday, 7 on Wednesday, and 2 on Thursday. He didn't get 3 pages done. How many pages of homework did his teachers give him?

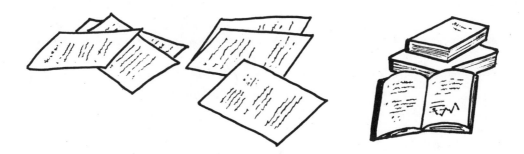

CREATE YOUR OWN

Eddy Phie's work is shown below. Create a story for each.

Eddy's Work	Your Own Problem
1. $8\overline{)680}$ = 85 My speed should be 85 kilometres per hour.	
2. $\begin{array}{r} 150.2 \\ -146.4 \\ \hline 3.8 \end{array}$ The difference is 3.8 centimetres.	
3. $\begin{array}{r} \$3.98 \\ \times\ \ 5 \\ \hline \$19.90 \end{array}$	
4. $\begin{array}{r} 14\frac{1}{4} \\ -\ 8\frac{3}{4} \\ \hline 5\frac{1}{2} \end{array}$ The length is $5\frac{1}{2}$ inches.	
5. $12\overline{)288}$ = 24 dozen	
6. $\begin{array}{r} \$8.75 \\ \times\ \ 14 \\ \hline \$122.50 \end{array}$ My wages will be \$122.50.	
7. $\begin{array}{r} 14.8 \\ 32.3 \\ 26.5 \\ \hline 73.6 \end{array}$ The distance is 73.6 units.	

<u>Create</u> <u>Your</u> <u>Own</u>

Mathematics teaching objective:

. Create single and multiple step word problems.

Problem-solving skills pupils <u>might</u> use:

. Search printed material for needed information.

. Create a problem which can be solved by certain solution procedures.

Materials needed:

. None

Comments and suggestions:

. Pupils are often asked to translate real life situations into math symbols. Here is an opportunity to do the reverse. Again, the focus is on understanding the problem.

. The activity provides each pupil a chance to express his/her creativity. Some will create the minimum amount while others will produce an intricate problem often involving unnecessary information.

. Problem 8 involves an average weight which may need to be explained. For those pupils not understanding the concept of average, you could use this problem as an example:

Five large bags of potatoes hold 160, 185, 182, 178 and 170 pounds of potatoes. Sally has the job of moving potatoes from bag to bag until each bag has the same number of pounds of potatoes. How many pounds will each bag have?

Answers:

Answers will vary.

Create Your Own (cont.)

Eddy's Work	Your Story Problem
8. 14 points 17 points 23 points ————— 54 points $3\overline{)54}$ with 18 on top The average is 18 points.	
9. $(4 \times \$.89) + (2 \times \$4.95) =$ $\$13.46$ The presents will cost $\$13.46$.	
10. 2 hours 15 min. 3 hours 40 min. 3 hours 55 min. 2 hours 45 min. ————————— ~~10 hours 155 min.~~ 12 hours 35 min. The Total time was 12 hours 35 minutes.	

Grade 6

IV. FRACTIONS

IV. FRACTIONS

These activities introduce the concepts of naming, ordering, equivalent, improper, mixed, measuring with, and the four operations with fractions. Strips of paper are used as a region model for fractions which later provides for a transition to the number line model. Pupils first make their own fraction models by folding strips of paper. Later, fraction strips and number rays reproduced from masters included in these materials are used. The activities in this part are developmental and should be used in sequence. The above concepts are introduced using a manipulative approach. No special vocabulary or formal algorithms are used. While using the activities, you will need to decide when and if your pupils are ready for the appropriate algorithm. Exercises from your textbook can supplement these materials and provide practice on the algorithms.

Whenever possible, word forms for fractions should be used. Pupils may better understand that 3 eighths + 2 eighths = 5 eighths rather than the more abstract form $\frac{3}{8} + \frac{2}{8} = \frac{5}{8}$.

DEVELOPING COMPUTATIONAL SKILLS, the 1978 Yearbook of the National Council of Teachers of Mathematics, describes two demonstrations to introduce pupils to the following important concepts.

1. The size of the unit. Display two differently-sized candy bars. Ask pupils to choose which one they would take. Stress that the size of the unit is important.

2. Equal-sized pieces. Display two candy bars divided into three pieces each as shown.

    Pupils recognize that a "fair" share is possible only if the pieces of the candy bars are the same size.

3. The number of pieces. Display two candy bars divided as shown.

    Pupils recognize that knowing the number of pieces the candy bar is divided into determines the size of each piece--more pieces-smaller size versus fewer pieces-larger size.

FOLDING FRACTIONS

Which is larger, $\frac{3}{4}$ of a strip of tagboard or $\frac{8}{12}$ of the same strip?

1. You can check the above question by getting strips of tagboard from your teacher.

   a. Use one strip. Label it 1 whole.

   b. Use another strip. Fold and crease it into 2 equal parts. Label each part $\frac{1}{2}$.

   c. Use another strip. Fold it into 2 equal parts. Now fold each of these into 2 equal parts, so you have 4 equal parts. Label each part $\frac{1}{4}$.

   d. Use the same method to fold and crease a strip into 8 equal parts. Label each part $\frac{1}{8}$.

   e. Carefully fold a strip into 3 equal parts. (<u>Do not crease</u> the folds until you are sure the parts are equal.) Label each part $\frac{1}{3}$.

   f. Fold a strip into 3 equal parts. Fold each of these into 2 equal parts, so you have 6 equal parts. Label each part $\frac{1}{6}$.

   g. Use the same method to fold a strip into 12 equal parts. Label each part $\frac{1}{12}$.

Folding Fractions

Mathematics teaching objectives:

. Establish need to have a unit region for comparing fractional parts.

. Name fractional parts of a region.

. Compare fractional parts of regions.

Problem-solving skills pupils might use:

. Make and/or use a model.

. Study the solution process.

Materials needed:

. 7 strips of tagboard for each pupil. (See page titled "Fraction Strips.")

Comments and suggestions:

. The first part of this activity should be directed by the teacher. Have pupils label a strip with 1 whole. Have pupils fold a second strip into 2 equal parts. Leading questions to ask are, "How many equal parts do you have?" "What would we call each equal part?" Have pupils label one part with the word name one-half and the other with the fraction name $\frac{1}{2}$. Similarly, fold and label strips for fourths and eighths. Fold and label strips for thirds, sixths, and twelfths. Warn pupils of the difficulty of folding thirds. Help pupils with spelling when they write the word names.

. For each strip have pupils (1) touch and name the first equal part, (2) start at the left end and slide a finger through two equal parts and name $\frac{2}{3}$, $\frac{2}{4}$, etc., (3) slide and name each of the other fractional parts.

. After this teacher-led folding, have pupils finish the remainder of the activity individually or with a partner.

Answers:

2. $\frac{3}{4}$ is larger than $\frac{8}{12}$.

3. a. $\frac{1}{2}$ b. $\frac{1}{4}$ c. $\frac{3}{6}$ d. $\frac{1}{2}$ e. 1

5. a. $\frac{1}{5}$ b. $\frac{3}{4}$ c. 1   d. $\frac{2}{7}$ e. 1   f. $\frac{7}{8}$

Folding Fractions (cont.)

2. Now which is larger, $\frac{3}{4}$ or $\frac{8}{12}$ of a strip? _____

3. Use the strips to help you. Circle the <u>larger</u> part.

   a. $\frac{1}{2}$ or $\frac{1}{3}$ b. $\frac{1}{6}$ or $\frac{1}{4}$ c. $\frac{2}{6}$ or $\frac{3}{6}$

   d. $\frac{5}{12}$ or $\frac{1}{2}$ e. 1 or $\frac{7}{8}$

4. Place your strips in order by size of one part. Do it from largest to smallest.

5. Suppose you had three other strips folded into five equal parts, seven equal parts, and nine equal parts. Circle the <u>larger</u> part.

   a. $\frac{1}{7}$ or $\frac{1}{5}$ b. $\frac{3}{4}$ or $\frac{3}{5}$ c. $\frac{8}{9}$ or 1

   d. $\frac{2}{7}$ or $\frac{2}{9}$ e. $\frac{5}{7}$ or 1   f. $\frac{7}{9}$ or $\frac{7}{8}$

FRACTION STRIPS

MARKED FRACTION STRIPS

ONE WHOLE	1											
ONE-HALF	$\frac{1}{2}$						$\frac{1}{2}$					
ONE-THIRD	$\frac{1}{3}$				$\frac{1}{3}$				$\frac{1}{3}$			
ONE-FOURTH	$\frac{1}{4}$			$\frac{1}{4}$			$\frac{1}{4}$			$\frac{1}{4}$		
ONE-SIXTH	$\frac{1}{6}$		$\frac{1}{6}$		$\frac{1}{6}$		$\frac{1}{6}$		$\frac{1}{6}$		$\frac{1}{6}$	
ONE-EIGHTH	$\frac{1}{8}$	$\frac{1}{8}$		$\frac{1}{8}$	$\frac{1}{8}$		$\frac{1}{8}$	$\frac{1}{8}$		$\frac{1}{8}$	$\frac{1}{8}$	
ONE-TWELFTH	$\frac{1}{12}$	$\frac{1}{12}$	$\frac{1}{12}$	$\frac{1}{12}$	$\frac{1}{12}$	$\frac{1}{12}$	$\frac{1}{12}$	$\frac{1}{12}$	$\frac{1}{12}$	$\frac{1}{12}$	$\frac{1}{12}$	$\frac{1}{12}$

MATCHING FRACTIONS

The fraction strips can show exact matches.

Example: Use the $\frac{1}{4}$-strip and the $\frac{1}{8}$-strip.   Match them by laying one on top of the other.

$\frac{1}{4}$		$\frac{1}{4}$		$\frac{1}{4}$		$\frac{1}{4}$	
$\frac{1}{8}$	$\frac{1}{8}$	$\frac{1}{8}$	$\frac{1}{8}$	$\frac{1}{8}$	$\frac{1}{8}$	$\frac{1}{8}$	$\frac{1}{8}$

This shows that $\frac{2}{8}$ matches $\frac{1}{4}$ or $\frac{2}{8} = \frac{1}{4}$.

Also, $\frac{4}{8}$ matches $\frac{2}{4}$ or $\frac{4}{8} = \frac{2}{4}$.

And $\frac{6}{8}$ matches $\frac{3}{4}$ or $\frac{6}{8} = \frac{3}{4}$.

Finally, $\frac{8}{8}$ matches $\frac{4}{4}$ or $\frac{8}{8} = \frac{4}{4}$.

Use your fraction strips to show many exact matches.
Record your results below and on the back.

Matching Fractions

Mathematics teaching objectives:

. Find equivalent fractions.

Problem-solving skills pupils might use:

. Use a model.

. Study the solution process.

. Look for patterns.

Materials needed:

. Set of fraction strips. The marked fraction strips work best rather than the pupil folded strips.

Comments and suggestions:

. The fraction strips provide an easy way to show equivalent fractions. By matching strips together, pupils should see "families" of equivalent fractions.

. As a check for pupils, you might make a transparency of the page titled "Marked Fraction Strips." Laying a straight edge on the " $\frac{1}{2}$ - mark" shows that $\frac{1}{2} = \frac{2}{4} = \frac{3}{6} = \frac{4}{8} = \frac{6}{12}$.

Answers:

These equivalent fractions can be shown with the fraction strips.

$$\frac{1}{2} = \frac{2}{4} = \frac{3}{6} = \frac{4}{8} = \frac{6}{12} \qquad \frac{1}{3} = \frac{2}{6} = \frac{4}{12} \qquad \frac{1}{4} = \frac{2}{8} = \frac{3}{12} \qquad \frac{1}{6} = \frac{2}{12}$$

$$\frac{2}{3} = \frac{4}{6} = \frac{8}{12} \qquad \frac{3}{4} = \frac{6}{8} = \frac{9}{12} \qquad \frac{5}{6} = \frac{10}{12}$$

$$1 = \frac{2}{2} = \frac{3}{3} = \frac{4}{4} = \frac{6}{6} = \frac{8}{8} = \frac{12}{12}$$

NUMBER LINE FRACTIONS 1

1. Get two copies of the Number Line Fractions 2 page. Get your fraction strips.

2. Cut one page on the dotted line.

3. Match up the lines and tape the two pages together to make one long sheet.

4. Starting at the dotted line, how many times will the <u>whole</u> <u>fraction</u> <u>strip</u> fit on the top number line? _____

5. Mark and label these points on the top number line: 0, 1, and 2.

6. Use the $\frac{1}{2}$-strip to mark and label these points on the second number line: $\frac{0}{2}, \frac{1}{2}, \frac{2}{2}, \frac{3}{2}, \frac{4}{2}$.

7. Use the other fraction strips to label the other number lines. Your $\frac{1}{8}$-line should look like this.

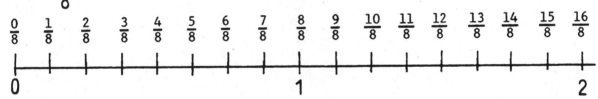

8. Use your completed number lines. Write three different names for

    a. 0 ___ ___ ___   b. 1 ___ ___ ___   c. 2 ___ ___ ___

9. Circle the larger fraction of each pair.

    a. $\frac{11}{8}$ or $\frac{12}{8}$ b. $\frac{5}{3}$ or $\frac{5}{4}$ c. $\frac{3}{2}$ or $\frac{5}{4}$ d. $\frac{15}{8}$ or $\frac{10}{6}$

10. Suppose you had other number lines marked with fifths, sevenths, and elevenths. Circle the larger fraction of each pair.

    a. $\frac{8}{8}$ or $\frac{8}{7}$ b. $\frac{13}{11}$ or $\frac{13}{12}$ c. $\frac{9}{5}$ or $\frac{9}{7}$ d. $\frac{6}{3}$ or $\frac{15}{7}$

Number Line Fractions 1

Mathematics teaching objectives:

. Make a transition from a region model to a number line model.

. Compare fractions greater than 1.

. Write equivalent fractions for 0, 1 and 2.

Problem-solving skills pupils might use:

. Use a model.

. Study the solution process.

Materials needed:

. Set of fraction strips. The marked fraction strips should be used.

. Two copies of "Number Line Fractions 2."

. Tape

Comments and suggestions:

. Several examples, perhaps with the overhead projector, should be done with pupils. Be sure they carefully locate 0, 1 and 2.

. The finished Number Line Fraction chart is useful for comparing fractions and for finding equivalent fractions.

. Pupils should save their completed Number Line Fraction charts for use in later activities.

Answers:

4. Two times

8. a. $0 = \frac{0}{2} = \frac{0}{3} = \frac{0}{4} = \frac{0}{6} = \frac{0}{8} = \frac{0}{12}$

    b. $1 = \frac{2}{2} = \frac{3}{3} = \frac{4}{4} = \frac{6}{6} = \frac{8}{8} = \frac{12}{12}$

    c. $2 = \frac{4}{2} = \frac{6}{3} = \frac{8}{4} = \frac{12}{6} = \frac{16}{8} = \frac{24}{12}$

9. a. $\frac{12}{8}$ b. $\frac{5}{3}$ c. $\frac{3}{2}$ d. $\frac{15}{8}$

10. a. $\frac{8}{7}$ b. $\frac{13}{11}$ c. $\frac{9}{5}$ d. $\frac{15}{7}$

NUMBER LINE FRACTIONS 2

MIXED AND IMPROPER

How much longer than 1 is $\frac{17}{12}$?

How many $\frac{1}{6}$'s does it take to make $1\frac{4}{6}$?

1. Use your number line fractions. How much longer than 1 is

   a. $\frac{11}{8}$? ____   b. $\frac{19}{12}$? ____   c. $\frac{5}{4}$? ____   d. $\frac{5}{3}$? ____

2. Complete these statements.

   a. $\frac{13}{8} = 1\frac{}{8}$ b. $\frac{17}{12} = 1\frac{5}{}$ c. $\frac{7}{4} = 1-$ d. $\frac{3}{2} =$ ____

3. Use your number line fractions. How many

   a. $\frac{1}{6}$'s are needed to make $1\frac{4}{6}$? _____

   b. $\frac{1}{8}$'s are needed to make $1\frac{7}{8}$? _____

   c. $\frac{1}{12}$'s are needed to make 1 ? _____

   d. $\frac{1}{4}$'s are needed to make $2\frac{3}{4}$? _____

4. Complete these statements.

   a. $1\frac{5}{6} =$ _____-sixths   b. $1\frac{7}{8} = 15-$ _____

   c. $1\frac{4}{12} = \frac{}{12}$ d. $2 = \frac{}{3}$ e. $1\frac{3}{4} =$ ____

Mixed and Improper

Mathematics teaching objectives:

. Recognize fractions greater than 1.

. Change between mixed and improper fractions.

Problem-solving skills pupils might use:

. Use a model.

. Study the solution process.

Materials needed:

. Number Line Fractions chart.

Comments and suggestions:

. The purpose of the activity is to change between mixed and improper fractions without relying on the traditional algorithms. The number line chart should promote an understanding that $\frac{11}{8}$ is $\frac{3}{8}$ longer than 1 or $\frac{11}{8}$ is the same as $1\frac{3}{8}$. Also that there are seven $\frac{1}{6}$'s in $1\frac{1}{6}$ or $1\frac{1}{6} = \frac{7}{6}$.

Answers:

1. a. $\frac{3}{8}$ b. $\frac{7}{12}$ c. $\frac{1}{4}$ d. $\frac{2}{3}$

2. a. $\frac{13}{8} = 1\frac{5}{8}$ b. $\frac{17}{12} = 1\frac{5}{12}$ c. $\frac{7}{4} = 1\frac{3}{4}$ d. $\frac{3}{2} = 1\frac{1}{2}$

3. a. 10   b. 15   c. 12   d. 11

4. a. $1\frac{5}{6}$ = 11-sixths   b. $1\frac{7}{8}$ = 15-eighths

   c. $1\frac{4}{12} = \frac{16}{12}$ d. $2 = \frac{6}{3}$ e. $1\frac{3}{4} = \frac{7}{4}$

WHERE WILL FLOPPIE LAND?

1. For each number line, estimate where the circled number should be.

2. a. Check your estimates. Use a centimetre ruler or edge of a sheet of paper. Mark the point where the circled number is located. Label it.

   b. Were you able to make reasonable estimates? _____

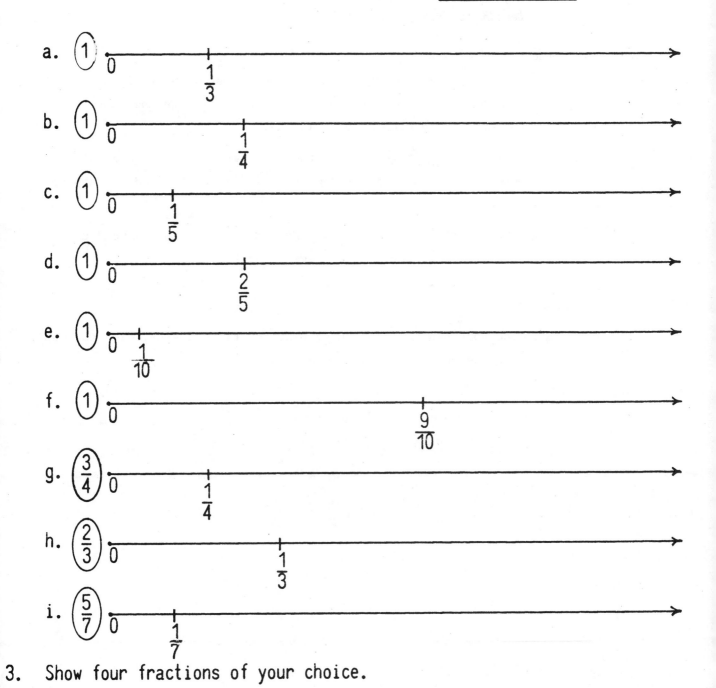

a. ① 0 ——————|—————————————————————→
                $\frac{1}{3}$

b. ① 0 ——————————————|———————————————→
                       $\frac{1}{4}$

c. ① 0 ————|——————————————————————————→
           $\frac{1}{5}$

d. ① 0 ——————————————|———————————————→
                      $\frac{2}{5}$

e. ① 0 —|————————————————————————————→
        $\frac{1}{10}$

f. ① 0 ——————————————————————————|————→
                                  $\frac{9}{10}$

g. ③⁄₄ 0 ——————|—————————————————————→
               $\frac{1}{4}$

h. ②⁄₃ 0 ————————————|———————————————→
                      $\frac{1}{3}$

i. ⑤⁄₇ 0 ————|——————————————————————————→
             $\frac{1}{7}$

3. Show four fractions of your choice.

   0 ————————————————————————————————|————→
                                      1

Where Will Floppie Land?

Mathematics teaching objectives:
- Locate points on a number line.
- Develop intuitive concepts of measurement.

Problem-solving skills pupils might use:
- Make and/or use a drawing.
- Make reasonable estimates.

Materials needed:
- Centimetre ruler (optional)

Comments and suggestions:
- Be sure pupils understand the numbers listed on the left side of the page are the ones to be shown. Before measuring, have pupils estimate where each number will be.

- An excellent alternative to the ruler is to use the edge of a note card. Make a point and use this length to find the appropriate number.

- It's important for pupils to know that two parts are needed--zero (the starting place) and a given length.

Answers:

- Make your own key on tracing paper or transparency. You can easily check answers by laying the key on the pupils' papers.

ADDING FRACTIONS

1. Work with a partner so you have two sets of fraction strips.

2. Suppose you placed $\frac{1}{2}$ and $\frac{1}{4}$ together. Use other strips to find three different fraction answers for the sum of $\frac{1}{2} + \frac{1}{4}$. Record.

   a. $\frac{1}{2} + \frac{1}{4} =$ b. $\frac{1}{2} + \frac{1}{4} =$ c. $\frac{1}{2} + \frac{1}{4} =$

3. Use your fraction strips to find these fraction sums.

   a. $\frac{3}{8} + \frac{1}{2} =$ b. $\frac{1}{2} + \frac{1}{3} =$

   c. $\frac{1}{3} + \frac{1}{6} =$ d. $\frac{3}{8} + \frac{1}{4} =$

   e. $\frac{3}{4} + \frac{3}{8} =$ f. $\frac{1}{3} + \frac{5}{6} =$

   g. $\frac{3}{4} + \frac{5}{12} =$ h. $\frac{7}{12} + \frac{2}{3} =$

4. Use your fraction strips to find these fraction sums.

   a. $\frac{1}{2} + \frac{1}{3} + \frac{1}{6} =$ b. $\frac{1}{2} + \frac{1}{3} + \frac{1}{4} =$

   c. $\frac{1}{3} + \frac{1}{6} + \frac{1}{12} =$ d. $\frac{1}{2} + \frac{1}{3} + \frac{1}{4} + \frac{1}{12} =$

EXTENSION

$$\frac{1}{2} + \frac{1}{3} + \frac{1}{4} + \frac{1}{6} + \frac{1}{8} + \frac{1}{12} =$$

© PSM 81

Adding Fractions

Mathematics teaching objectives:

    . Add fractions with unlike denominators.

Problem-solving skills pupils <u>might</u> use:

    . Use a model.
    . Make reasonable estimates.
    . Look for patterns.

Materials needed:

    . Marked fraction strips.

Comments and suggestions:

    . A teacher demonstration for the first problem may be necessary. The activity provides a way of adding fractions without having to change to common denominators.

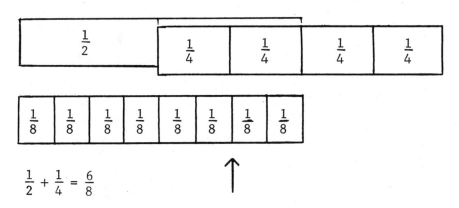

$$\frac{1}{2} + \frac{1}{4} = \frac{6}{8}$$

    . Since this packet of materials has not stressed reducing fractions to lowest terms, pupils may find several equivalent answers as they work with the fraction strips.

Answers:

2.   a. $\frac{1}{2} + \frac{1}{4} = \frac{3}{4}$ b. $\frac{1}{2} + \frac{1}{4} = \frac{6}{8}$ c. $\frac{1}{2} + \frac{1}{4} = \frac{9}{12}$

3.   a. $\frac{7}{8}$ b. $\frac{5}{6}$ or $\frac{10}{12}$ c. $\frac{3}{6}$ or $\frac{1}{2}$ or $\frac{6}{12}$ or $\frac{4}{8}$

     d. $\frac{5}{8}$ e. $\frac{9}{8}$ f. $\frac{7}{6}$ or $\frac{14}{12}$ g. $\frac{14}{12}$ or $\frac{7}{6}$ h. $\frac{15}{12}$

4.   a. $\frac{6}{6}$ b. $\frac{13}{12}$ c. $\frac{7}{12}$ d. $\frac{14}{12}$ or $\frac{7}{6}$

5. $\frac{35}{24}$

HOW MUCH LONGER?

Use your fraction strips.

1. a. $\frac{3}{4}$ is how much longer than $\frac{1}{4}$? _____

    b. $\frac{5}{6}$ is how much longer than $\frac{4}{6}$? _____

    c. $\frac{7}{8}$ is how much longer than $\frac{2}{8}$? _____

    d. $\frac{5}{12}$ is how much longer than $\frac{3}{12}$? _____

    e. $\frac{12}{12}$ is how much longer than $\frac{3}{12}$? _____

2. a. $\frac{3}{4}$ is how much longer than $\frac{1}{2}$? _____

    b. $\frac{5}{6}$ is how much longer than $\frac{1}{3}$? _____

    c. $\frac{6}{8}$ is how much longer than $\frac{2}{4}$? _____

    d. $\frac{9}{12}$ is how much longer than $\frac{2}{3}$? _____

    e. $\frac{10}{12}$ is how much longer than $\frac{3}{4}$? _____

PSM 81

How Much Longer?

Mathematicts teaching objectives:
  . Subtract fractions.

Problem-solving skills pupils <u>might</u> use:
  . Use a model.
  . Make reasonable estimates.
  . Look for patterns.

Materials needed:
  . Marked fraction strips

Comments and suggestions:
  . This activity uses a comparative or additive model for subtraction.
    Pupils determine how much more is needed to make the shorter length
    the same as the longer length. A teacher demonstration may be helpful.

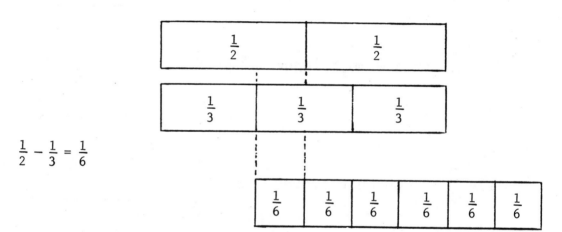

$$\frac{1}{2} - \frac{1}{3} = \frac{1}{6}$$

  . Since this packet has not stressed reducing fractions to lowest terms,
    pupils may find several equivalent answers as they work with the frac-
    tion strips.

Answers:

1. a. $\frac{2}{4}$ b. $\frac{1}{6}$ c. $\frac{5}{8}$ d. $\frac{2}{12}$ e. $\frac{9}{12}$

2. a. $\frac{1}{4}$ b. $\frac{3}{6}$ c. $\frac{2}{8}$ d. $\frac{1}{12}$ e. $\frac{1}{12}$

3. a. $\frac{1}{6}$ b. $\frac{1}{12}$ c. $\frac{1}{6}$ d. $\frac{1}{12}$ e. $\frac{7}{12}$

4. a. $\frac{3}{7}$ b. $\frac{3}{10}$ c. $\frac{2}{15}$

How Much Longer? (cont.)

3. a. $\frac{1}{2}$ is how much longer than $\frac{1}{3}$? _____

    b. $\frac{1}{3}$ is how much longer than $\frac{1}{4}$? _____

    c. $\frac{2}{3}$ is how much longer than $\frac{1}{2}$? _____

    d. $\frac{5}{6}$ is how much longer than $\frac{3}{4}$? _____

    e. $\frac{3}{4}$ is how much longer than $\frac{1}{6}$? _____

4. Suppose you had other fraction strips.

    a. $\frac{5}{7}$ is how much longer than $\frac{2}{7}$? _____

    b. $\frac{9}{10}$ is how much longer than $\frac{3}{5}$? _____

    c. $\frac{4}{5}$ is how much longer than $\frac{2}{3}$? _____

PSM 81

ADDING ON THE NUMBER LINE

1. Work with a partner so you have two sets of fraction strips. Use your fraction number lines also.

2. Find these sums by placing the strips along the indicated number line.

   a. $\frac{1}{8}$-line, $\frac{1}{8} + \frac{3}{8} =$ _____

   b. $\frac{1}{4}$-line, $\frac{3}{4} + \frac{3}{4} =$ _____

   c. $\frac{1}{12}$-line, $\frac{5}{12} + \frac{7}{12} =$ _____

   d. $\frac{1}{6}$-line, $\frac{1}{6} + \frac{3}{6} =$ _____

3. Find these sums by placing the strips along the indicated number line.

   a. $\frac{1}{4}$-line, $\frac{1}{2} + \frac{1}{4} =$ _____

   b. $\frac{1}{6}$-line, $\frac{2}{6} + \frac{1}{3} =$ _____

   c. $\frac{1}{8}$-line, $\frac{2}{4} + \frac{3}{8} =$ _____

   d. $\frac{1}{12}$-line, $\frac{3}{12} + \frac{5}{6} =$ _____

4. Write the number line you would use to find these sums.

   a. $\frac{1}{2} + \frac{1}{8}$ _____

   b. $\frac{2}{4} + \frac{5}{12}$ _____

   c. $\frac{2}{3} + \frac{3}{12}$ _____

   d. $\frac{1}{2} + \frac{3}{6}$ _____

Adding On The Number Line

Mathematics teaching objectives:

. Add fractions.

. Choose a common denominator.

. Find equivalent fractions.

Problem-solving skills pupils **might** use:

. Use a model.

. Solve an easier but related problem.

. Study the solution process.

Materials needed:

. Marked fraction strips

. Number Line Fractions chart

Comments and suggestions:

. This activity leads towards the standard algorithm for adding fractions. You can demonstrate this after pupils have completed the activity. For example, using the $\frac{1}{4}$ - line, $\frac{1}{2} + \frac{1}{4}$ becomes $\frac{2}{4} + \frac{1}{4}$ or $\frac{3}{4}$. Using the $\frac{1}{6}$ - line, $\frac{1}{2} + \frac{1}{3}$ becomes $\frac{3}{6} + \frac{2}{6} = \frac{5}{6}$.

. Problems 4 and 6 only ask that pupils determine the fraction number line that would work best. You may have pupils complete the problems.

Answers:

2. a. $\frac{4}{8}$ b. $\frac{6}{4}$ c. $\frac{12}{12}$ d. $\frac{4}{6}$

3. a. $\frac{3}{4}$ b. $\frac{4}{6}$ c. $\frac{7}{8}$ d. $\frac{13}{12}$

4. a. $\frac{1}{8}$ - line   b. $\frac{1}{12}$-line   c. $\frac{1}{12}$-line   d. $\frac{1}{6}$-line or $\frac{1}{12}$ - line

5. a. $\frac{5}{6}$ b. $\frac{10}{12}$ c. $\frac{9}{6}$ d. $\frac{17}{12}$

6. Answers may vary.

a. $\frac{1}{10}$-line   b. $\frac{1}{15}$-line   c. $\frac{1}{16}$-line

d. $\frac{1}{10}$-line   e. $\frac{1}{15}$-line   f. $\frac{1}{35}$-line

Adding on the Number Line (cont.)

5. Choose the best number line to use. Find these sums.

   a. $\dfrac{1}{2} + \dfrac{1}{3} =$ _____

   b. $\dfrac{1}{3} + \dfrac{2}{4} =$ _____

   c. $\dfrac{2}{3} + \dfrac{5}{6} =$ _____

   d. $\dfrac{2}{3} + \dfrac{3}{4} =$ _____

6. Suppose you had many other fraction number lines.
   Write the number line you would use to find these sums.

   a. $\dfrac{1}{2} + \dfrac{3}{10}$ ____

   b. $\dfrac{2}{3} + \dfrac{4}{15}$ ____

   c. $\dfrac{5}{16} + \dfrac{3}{8}$ ____

   d. $\dfrac{3}{5} + \dfrac{1}{2}$ ____

   e. $\dfrac{2}{3} + \dfrac{1}{5}$ ____

   f. $\dfrac{1}{5} + \dfrac{3}{7}$ ____

LENGTHS

1. The drawing shows that 3 lengths of $\frac{3}{8}$ have been traced.

   How many eighths in all have been traced? _____

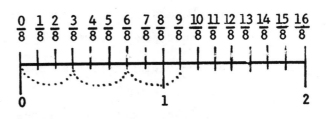

2. The drawing shows that 4 lengths of $\frac{1}{8}$ have been traced.

   How many eighths in all have been traced? _____

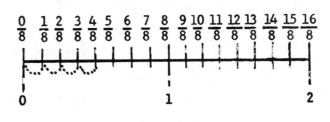

3. Use your fraction number lines. Trace out each problem. Record your answers.

   a. 3 lengths of $\frac{2}{8}$ _____

   b. 2 lengths of $\frac{1}{3}$ _____

   c. 5 lengths of $\frac{3}{12}$ _____

   d. 4 lengths of $\frac{3}{8}$ _____

   e. 3 lengths of $\frac{1}{2}$ _____

   f. 4 lengths of $\frac{2}{6}$ _____

   g. 3 lengths of $\frac{5}{12}$ _____

   h. 2 lengths of $\frac{7}{8}$ _____

4. Suppose you had other number lines marked in sevenths, ninths, and tenths. Find these answers.

   a. 3 lengths of $\frac{2}{7}$ _____

   b. 2 lengths of $\frac{3}{9}$ _____

   c. 6 lengths of $\frac{2}{10}$ _____

   d. 12 lengths of $\frac{4}{7}$ _____

Lengths

Mathematics teaching objectives:

    . Understand the concept of multiplying fractions by whole numbers.

    . Multiply a fraction by a whole number.

Problem-solving skills a pupil **might** use:

    . Use a model.

    . Study the solution process.

Materials needed:

    . Number Line Fractions chart.

Comments and suggestions:

    . This activity emphasizes the "groups of" approach to multiplication
That is, 3 x 2-eighths means 3 groups of (in this case, lengths of)
2-eighths. 2 x 3-eighths means 2 groups of 3-eighths. Although the
answer in each case is 6-eighths, the physical representation is
different.

    . The use of word names helps eliminate the tendency to say 3 groups
of $\frac{2}{8}$ is $\frac{6}{24}$.

Answers:

1. 9-eighths

2. 4-eighths

3. a. $\frac{6}{8}$ b. $\frac{2}{3}$ c. $\frac{15}{12}$ d. $\frac{12}{8}$

    e. $\frac{3}{2}$ f. $\frac{8}{6}$ g. $\frac{15}{12}$ h. $\frac{14}{8}$

4. a. $\frac{6}{7}$ b. $\frac{6}{9}$ c. $\frac{12}{10}$ d. $\frac{48}{7}$

MORE LENGTHS

1. a. On the $\frac{1}{6}$-line trace, with your finger, a length of $\frac{4}{6}$

   b. Trace $\frac{1}{2}$ of a length of $\frac{4}{6}$.

   c. $\frac{1}{2}$ of $\frac{4}{6}$ = _____

2. a. On the $\frac{1}{12}$-line trace a length of $\frac{8}{12}$.

   b. Trace $\frac{1}{4}$ of a length of $\frac{8}{12}$.

   c. $\frac{1}{4}$ of $\frac{8}{12}$ = _____

3. a. On the $\frac{1}{12}$-line trace a length of $\frac{9}{12}$.

   b. Trace $\frac{2}{3}$ of a length of $\frac{9}{12}$.

   c. $\frac{2}{3}$ of $\frac{9}{12}$ = _____

4. Use your fraction number lines. Trace out each problem. Record your answer.

   a. $\frac{1}{3}$ of a length of $\frac{3}{8}$ _____   b. $\frac{1}{4}$ of a length of $\frac{4}{6}$ _____

   c. $\frac{2}{3}$ of a length of $\frac{6}{8}$ _____   d. $\frac{3}{4}$ of $\frac{8}{12}$ _____

   e. $\frac{1}{2}$ of $\frac{6}{8}$ _____

More Lengths

Mathematics teaching objectives:

. Understand the concept of multiplication of fractions.

. Multiply a fraction by a fraction (readiness).

Problem-solving skills pupils might use:

. Use a model.

. Study the solution process.

Materials needed:

. Number Line Fractions chart.

Comments and suggestions:

. Pupils need this understanding of fractions:

Two-thirds of a length means to separate the length into three equal lengths and then use two of the equal lengths. For example, $\frac{2}{3}$ of a length of $\frac{6}{8}$ means to separate the $\frac{6}{8}$ into three equal lengths of $\frac{2}{8}$ each and then use two of the lengths to get $\frac{4}{8}$.

. This activity may need teacher direction.

Answers:

1. c. $\frac{2}{6}$

2. c. $\frac{2}{12}$

3. c. $\frac{6}{12}$

4. a. $\frac{1}{8}$ b. $\frac{1}{6}$ c. $\frac{4}{8}$ d. $\frac{6}{12}$ e. $\frac{3}{8}$

5. a. $\frac{1}{4}$ b. $\frac{1}{3}$ c. $\frac{3}{8}$

More Lengths (cont.)

5. Trace out each problem on the $\frac{1}{2}$-line. Find your answer on one of the other fraction number lines.

    a. $\frac{1}{2}$ of a length of $\frac{1}{2}$ _____

    b. $\frac{2}{3}$ of a length of $\frac{1}{2}$ _____

    c. $\frac{3}{4}$ of a length of $\frac{1}{2}$ _____

EQUAL LENGTHS

1. A length of $\frac{8}{12}$ is cut to show 4 equal lengths.
   How long is each length? _____

2. A length of $\frac{10}{12}$ is cut into smaller lengths of $\frac{2}{12}$.
   How many are cut? _____

3. How long is the length if 3 smaller lengths of $\frac{2}{8}$ can be cut
   from it? _____

4. A length of $\frac{15}{12}$ is divided into smaller lengths of $\frac{3}{12}$.
   How many pieces are there? _____

5. A length of $\frac{12}{8}$ is divided to show 3 equal lengths.
   How long is each length? _____

6. How long is the length if 5 smaller lengths of $\frac{3}{8}$ can be cut
   from it? _____

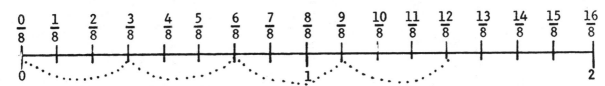

This diagram shows a length of $\frac{12}{8}$ marked in smaller lengths of $\frac{3}{8}$.
It also shows a length of $\frac{12}{8}$ marked into 4 equal lengths. These
are shown by division statements.

$$\frac{12}{8} \div \frac{3}{8} = 4 \qquad \text{or} \qquad \frac{12}{8} \div 4 = \frac{3}{8}$$

7. Describe this diagram. Write two division statements about the
   diagram.

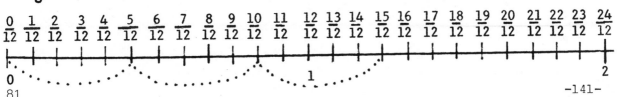

Equal Lengths

Mathematics teaching objectives:

. Understand the concept of division with fractions.

. Divide a fraction by another fraction or by a whole number.

Problem-solving skills pupils might use:

. Use a model.

. Study the solution process.

Materials needed:

. Number Line Fractions chart

Comments and suggestions:

. The operation of division is the hardest for pupils to do. The concept of division is much easier. These division problems ask either how many lengths there are or how large is each group. For example, $\frac{6}{8} \div 3$ means how long is each length if $\frac{6}{8}$ is divided into 3 equal lengths. And $\frac{6}{8} \div \frac{2}{8}$ means how many lengths are there if $\frac{6}{8}$ is divided into equal lengths of $\frac{2}{8}$ each.

. To emphasize both of the above meanings, this activity might work best as a teacher directed activity.

. This interpretation is very difficult using a problem like $\frac{1}{2} \div \frac{2}{3}$. It is best to make the explanation with easily understood problems.

Answers:

1. $\frac{2}{12}$ 2. 5   3. $\frac{6}{8}$

4. 5   5. $\frac{4}{8}$ 6. $\frac{15}{8}$

7. $\frac{15}{12} \div \frac{5}{12} = 3$ or $\frac{15}{12} \div 3 = \frac{5}{12}$

Grade 6

V. GEOMETRY

V.   GEOMETRY

Children have a natural interest in geometric ideas because they live in a world filled with shapes and solids. The middle school years provide an excellent opportunity to continue the study of space relationships and the attributes of geometric objects. Fortunately this study can also be used to improve the ability of students to visualize, to allow the "number poor" but "geometric rich" student to shine and to emphasize problem solving skills.

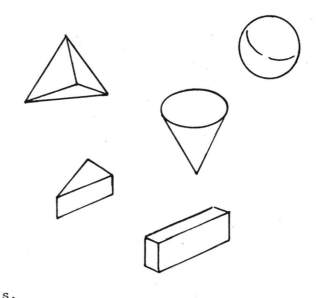

Using the Activities

Because curriculum guides seldom agree on how and when the topics of geometry should be taught in the middle grades, it is a good idea to coordinate your geometry lessons with fifth- and sixth-grade teachers. The problem solving sections could be shared with some lessons switched to different grade levels as appropriate for your curriculum guide.

The activities can be used when related topics are covered in the text-book or they may be used at any time during the year. Possible problems students might have with the geometry are discussed on each page. Many of the activities are meant for small groups or partners. An activity can be done by one group while the rest are doing other work at their desks. Some activities might be used at interest stations for pupils with unassigned time. If you have enough material, several groups can do the activity at one time.

The materials needed for these activities include narrow strips of paper (or uncooked spaghetti), scissors, grid paper, small tiles or squares, geoboards and rubber bands, and cubes. These materials make the activities easier but many can be completed using only the pupil page.

TRIANGLES HAVE THREE SIDES

Needed: Pieces of uncooked spaghetti (or narrow strips of paper or
         Cuisenaire rods) that are 2, 3, 4, 5, 6, 7, 8, 9, & 10 cm long

1. Make at least ten triangles
   using the strips. Record
   the length of the sides in
   the table to the right.

Triangle	Longest Side	Shortest Side	Other Side
1			
2			
3			
4			
5			
6			
7			
8			
9			
10			

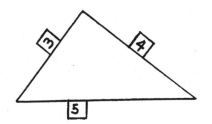

2. Which of these
   make a triangle?

   Use the strips
   to decide.

Triangle	Longest Side	Shortest Side	Other Side	Triangle Yes or No
11	10	4	5	
12	8	3	6	
13	7	2	3	
14	9	4	5	
15	10	2	6	

3. Study the two tables. For the strips to form a triangle,
   what has to be true?

4. Use your rule. Write the lengths of the sides for 5 more
   triangles you can make with the strips. Then make each with
   the strips to check yourself.

a._____   b._____   c._____   d._____   e._____

) PSM 81

Triangles Have Three Sides

Mathematics teaching objectives:

. Discover a property of triangles--the sum of the lengths of the
   two shorter sides is greater than the length of the longest side.

Problem-solving skills pupils _might_ use:

. Collect data needed to solve the problem.

. Make and use a model.

. Look for patterns and properties.

Materials needed:

. Narrow strips of paper (uncooked spaghetti works well) in lengths
   of 2, 3, 4, 5, 6, 7, 8, 9 and 10 cm.

Comments and suggestions:

. If strips of paper are used, have pupils make
   triangles by matching up corners so the triangle
   is on the inside. Watch for pupils who get
   "triangles" by not _exactly_ matching the corners,
   e.g. a "4-6-10" triangle.

. Label the strips for ease in recording. If
   spaghetti is used, label it with a small piece
   of tape.

Answers:

1. Answers will vary.

2. Triangle 11 - No, 12 - Yes, 13 - No, 14 - No, 15 - No

3. The sum of the lengths of the two shorter sides is greater
    than the length of the longest side.

4. Answer will vary but the condition in (3) must be observed.

TRIANGLES HAVE THREE ANGLES

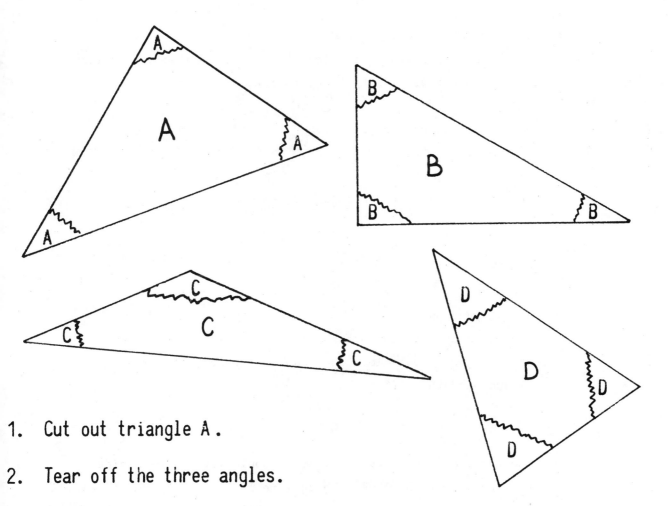

1. Cut out triangle A.

2. Tear off the three angles.

3. Fit the three angles, <u>side</u> <u>by</u> <u>side</u>, on the dot below.
    Start with one side on the line.

●————————————————————————●

4. Repeat, using triangles B, C, and D.

5. What did you discover? _____

6. Repeat with two triangles of your own.

Triangles Have Three Angles

Mathematics teaching objectives:

    . Discover a property of triangles - the three angles of a triangle
      will fit together, side-by-side, to make a straight line.

Problem-solving skills pupils <u>might</u> use:

    . Use a model.

    . Look for a property.

    . Make explanations based upon data.

Materials needed:

    . Scissors

Comments and suggestions:

    . If pupils decide to cut off the angles, instead of tearing,
      have them mark the angle. They can then place the marked
      angles around the dot.

Answers:

    5. The three angles of each triangle should fit, side-by-side,
        to make a straight line. To be more formal - the sum of
        the measures of the three angles of a triangle is 180°.

    6. The property holds for any triangle.

QUADRILATERALS HAVE FOUR ANGLES

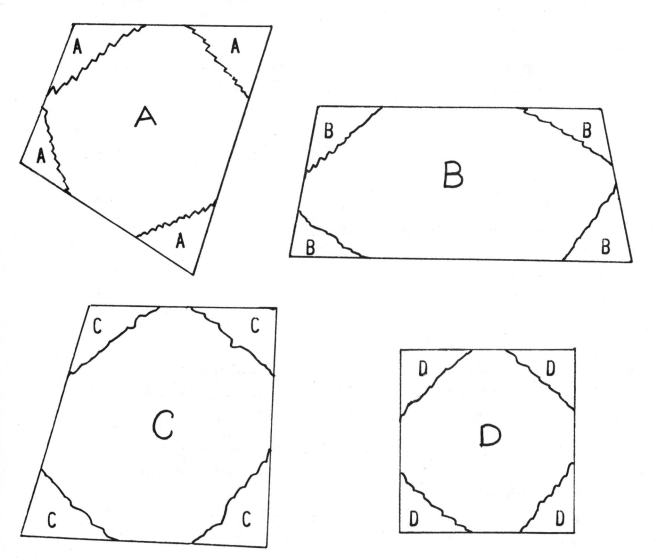

1. Cut out quadrilateral (four-sided) **A**.

2. Tear off the four corners.

3. Fit the four corners, <u>side</u> <u>by</u> <u>side</u>, on the dot below.
   Start with one side on the line.

4. Repeat, using quadrilaterals B, C, and D.
5. What did you discover? _____
6. What would happen with 5-sided and 6-sided figures?
   Experiment to check your prediction.

Quadrilaterials Have Four Angles

Mathematics teaching objectives:

    . Discover a property of quadrilaterals - the four angles of a
      quadrilateral will fit together, side-by-side, to completely
      fill the space around a dot.

Problem-solving skills pupils <u>might</u> use:

    . Use a model.

    . Look for a property.

    . Make explanations based upon data.

Materials needed:

    . Scissors

Comments and suggestions:

    . Some pupils may need review on the meaning of quadrilaterial -
      a four-sided polygon. Some common examples are squares, rec-
      tangles, parallelograms and trapezoids.

    . If pupils decide to cut off the angles, instead of tearing,
      have them mark the angle. They can then place the marked
      angles around the dot.

Answers:

5. The four angles of a quadrilateral should fit, side-by-side,
    to completely fill the space around a dot. More formally,
    the sum of the measures of the four angles of a quadrilateral
    is 360°.

6. Five-sided polygons - the sum is 540°, $1\frac{1}{2}$ times around the dot.

    Six-sided polygons - the sum is 720°, 2 times around the dot.

MAKING PENTOMINOES

Needed: 5 squares or tiles
        grid paper, scissors

A pentomino is made by putting
5 squares together side by side.

This is a pentomino.

These are not pentominoes.

1. Use the squares to make a pentomino.

2. Mark it on the grid paper and carefully cut it out.

3. Make as many other pentominoes as you can.
   Hint: You can find more than 10 pentominoes.

Two pentominoes are the same if they can be exactly
matched together by turning or flipping.

is the same as   by turning.

is the same as   by flipping.

PSM 81

Making Pentominoes

Mathematics teaching objectives:

. Recognize shapes.
. Construct pentominoes.

Problem-solving skills pupils might use:

. Record solution attempts.
. Make and use a model.
. Guess and check.
. Find another answer when more than one is possible.

Materials needed:

. Five squares or tiles per pupil
. Grid paper
. Scissors

Comments and suggestions:

. Construct a pentomino on the overhead projector. Show pupils a
non-example to be sure they understand what a pentomino is. It is
helpful to have a completed pentomino to show pupils that

    and   are the same if one is flipped over.

. The activity goes well with pupils working in pairs to discuss and
cooperate.

Answers:

Twelve pentominoes are possible. Each can be associated with
a letter of the alphabet. A somewhat organized scheme can be used to
organize the pentomino pieces.

Five in a row: I

Four in a row: L   R

Three in a row: P   C   V   T

     S   F   Z   X

Two in a row: W

ACTIVITIES WITH PENTOMINOES

1. Find the pentomino having the least perimeter.

2. Find the pentominoes that will fold up to make a box with no lid.

3. Which pentominoes have <u>line symmetry</u>? This means a piece can be flipped over a middle line and exactly fit on itself.

4. Which pentominoes have <u>rotational symmetry</u>? This means a piece can be turned top to bottom and exactly fit on itself.

5. Play the Pentominoes Grid Game.

    a. Get an 8 by 8 grid.

    b. Let each player pick 6 pentomino pieces from a total set of 12.

    c. Take turns placing the pieces on the grid.

    d. The last player to place a piece is the winner. No pieces can overlap and no pieces can lie partly off the grid.

    e. <u>Challenge</u>: Play the game so all 12 pieces do lie on the grid. This game is a tie.

Activities With Pentominoes

Mathematics teaching objectives:

    . Review perimeter and area concepts.

    . Find pentominoes with symmetry.

    . Construct rectangles with pentomino pieces.

Problem-solving skills pupils *might* use:

    . Classify objects.

    . Use a model.

    . Guess and check.

Materials needed:

    . A set of the twelve pentomino pieces.

Comments and suggestions:

    . The lesson is a collection of activities to do with pentominoes. A pupil is not necessarily expected to do all the activities. Problem 1 deals with perimeter. Problem 2 is a folding activity. Problems 3 and 4 deal with symmetry.. Problems 5-10 are concerned with covering or making various rectangles.

    . Problems 8 and 9 are rather difficult. They can be used as a continuing project for pupils to explore. At some point it may be helpful to provide an answer and ask pupils to use it as a basis for finding another answer. Or provide a partial answer and allow pupils to complete the solution.

    . In some activities, pieces might need to be turned over to fit the space.

Answers:

1.   has a perimeter of 10 units.
               All others have a perimeter of 12 units.

2.   will *not* form an open cube.

3. I, C, V, T, X, and W have line symmetry.

4. I, Z and X have rotational symmetry.

Activities with Pentominoes (cont.)

6. Stack all 12 pentominoes on top of one another.
    What is the smallest rectangle needed so all pieces would
    fit inside the rectangle?

7. Make this shape using two pentominoes. It can be done in
    more than one way.

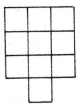

8. Find ways to put some pentominoes together to make rectangles.
    Record your solutions.

9. Use all 12 pentominoes to make a 6 by 10 rectangle.
    It is difficult to do but can be done in over 2000 ways.

10. Use all 12 pentominoes to make a 5 by 12 rectangle.

11. Use all 12 pentominoes to make a 4 by 15 rectangle.

12. Use all 12 pentominoes to make a 2 by 30 rectangle.

Activities With Pentominoes

Answers:

6. A 3 by 5 rectangle is needed.

7. Four ways are possible.

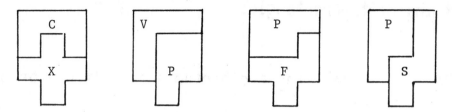

8. Using 2 pieces, rectangles are not possible. Many are possible using
    3 or more pieces. One rectangle, 3 by 5, can be made using pieces
    P, L, and V.   C, X, P, F, and L make a 5 by 5 rectangle. I, S, T,
    V, Z, W, R make a 5 by 7 rectangle.

9.

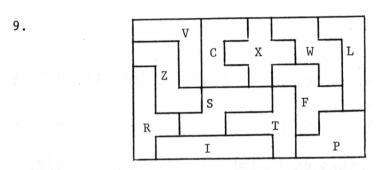

10.

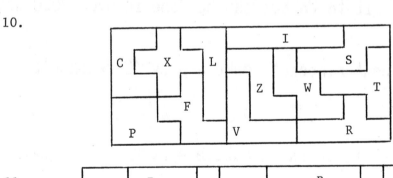

11.

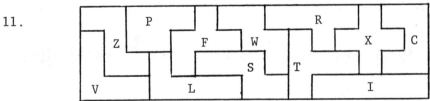

12. A 2 by 30 rectangle is impossible. Piece T is 3 units wide regardless
    of how it is positioned.

LINE INVESTIGATIONS ON A GEOBOARD

Materials needed: Geoboard
                     Rubber bands
                     Geoboard record paper

1. Make and record (if possible).
    a. two parallel lines
    b. two perpendicular lines
    c. one line that does not have a parallel
    d. one line that does not have a perpendicular
    e. two intersecting lines that are not perpendicular
    f. two lines that will intersect only if they are extended off the geoboard
    g. two lines that intersect at more than one point

2. Make two intersecting lines that together touch 9 nails. Repeat, using some different nails. Record all you can find.

3. Find and record as many lines of different length as you can. Hint: There are more than 12.

Line Investigations On A Geoboard

Mathematics teaching objectives:

. Review vocabulary of parallel, perpendicular and intersecting.
. Find different lengths of line segments.

Problem-solving skills pupils might use:

. Use a model.
. Identify a problem situation in which a solution is not possible.
. Record solution possibilities.
. Look for patterns.

Materials needed:

. Geoboard
. Rubber bands
. Geoboard record paper

Comments and suggestions:

. Using a geoboard allows pupils to try several possibilities and
to find a correct answer before recording. This avoids having
to erase mistakes resulting from incorrect beginnings.

Answers:

1. a, b, e and f have multiple answers. An example of each is shown;
c, d and g are impossible.

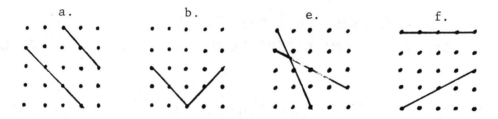

2. Several answers are possible.

3. Fourteen lines are possible. They are shown on five different record
sheets although all can be made using one corner nail on a geoboard.

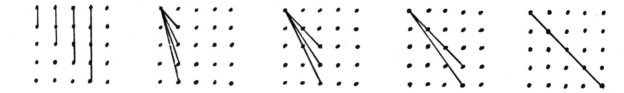

A MYSTERY FIGURE

Needed: Geoboard
           Rubber bands

       Detective Dee Tector had eliminated the robbery suspects
down to three persons--one from Moscow, another from Peking,
and the third from Los Angeles.   Ms. Tector found a geoboard
and the following instructions. With these clues she was able
to pinpoint the robber.

Clue 1: Make the longest line segment possible.

Clue 2: Make a line segment perpendicular to #1 that touches
         exactly 4 nails, one of which is a corner nail.

Clue 3: Make a line segment that (a) is parallel to #1,
         (b) intersects an endpoint of #2, and (c) touches
         exactly 3 nails.

Clue 4: Make a line segment perpendicular at an endpoint of
         #3 that touches exactly 2 nails.

Clue 5: Make a line segment that (a) touches the nail that
         both #2 and #3 touch, (b) touches only one nail of #2,
         (c) intersects one nail of #1, (d) touches exactly 3
         nails, and (e) does not intersect #4.

Clue 6: Make a line segment exactly 4 units long that connects
         2 corner nails.

Clue 7: Repeat Clue 6 for two other corner nails.

Clue 8: Repeat Clue 7 for two other corner nails.

Clue 9: Repeat Clue 8 for two other corner nails.

What you have made is a tangram shape. The tangram was invented
in China. Who is guilty? _____

A Mystery Figure

Mathematics teaching objectives:
    . Construct line segments.
    . Recognize shapes.

Problem-solving skills pupils might use:
    . Follow written directions.
    . Use a model.
    . Satisfy one condition at a time.

Materials needed:
    . Geoboard
    . Rubber bands

Comments and suggestions:
    . This answer is meaningful only if pupils have had experiences
      with tangrams. Clue 5 is the most difficult for pupils to
      understand. They may need a hint.

Answer:

. The person from Peking
  is guilty. The tangram
  is thought to have
  been invented in China.

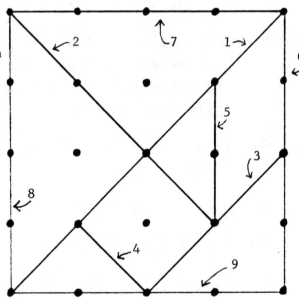

ALL THE SIDES

Materials needed: Cubes

1. Build each of these models. Find the surface area of each.
    Remember to include front, back, top, bottom, and both sides.

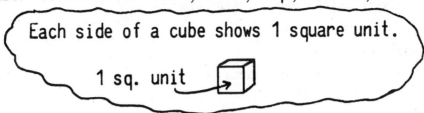

Each side of a cube shows 1 square unit.

1 sq. unit

a.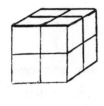

b.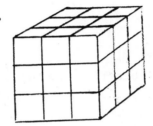

c.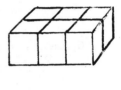

_____   _____   _____

d.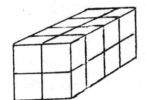

e.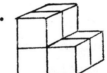

f.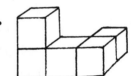

_____   _____   _____

2. Find the surface area of each model. Watch for hidden surfaces.

a.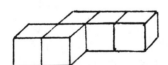

b.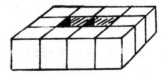

_____   _____

c.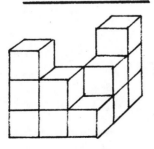

d.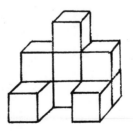

_____   _____

PSM 81

All The Sides

Mathematics teaching objectives:

   . Find surface area.

Problem-solving skills pupils **might** use:

   . Use a drawing or model.

   . Look for patterns.

   . Break a problem into parts.

Materials needed:

   . Cubes

Comments and suggestions:

   . Use of cubes will benefit some pupils. They may get down to
the same level as the model and look straight on at the front
or side, etc.

   . If enough cubes are not available, pupils can work in pairs.
Or this activity could be one station of a set of activities.

   . Pupils may determine that the areas occur in pairs. For
example, the surface area of the front will be the same as
back. In some models, all six surfaces will have the same area.

Answers:

   1. a. 24 sq. units   b. 54 sq. units   c. 22 sq. units
         d. 40 sq. units   e. 22 sq. units   f. 22 sq. units

   2. a. 22 sq. units   b. 40 sq. units
         c. 42 sq. units   d. 34 sq. units

SURFACE AREA PATTERNS WITH CUBES

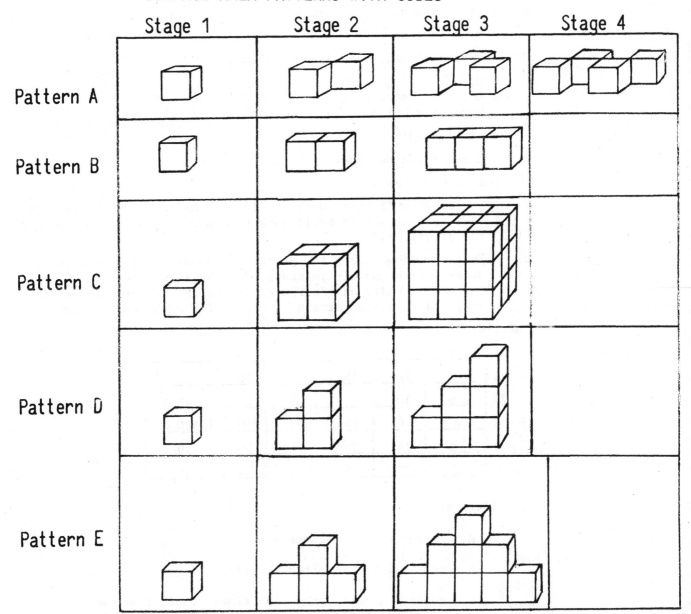

FOR EACH PATTERN: (a) Find the surface area in stages 1, 2, and 3.
                       (b) Predict the surface area for stages 4 and 5.
                       (c) Find the surface areas for stages 4 and 5 by drawing sketches
                             or making the models with cubes.
                       (d) Predict the surface area for stage 10.

Surface Area of Stage

Patterns	1	2	3	4	5	10
A						
B						
C						
D						
E						

-165-

PSM 81

Surface Area Patterns With Cubes

Mathematics teaching objectives:

. Find surface area.

Problem-solving skills pupils might use:

. Use a drawing or model.
. Look for patterns.
. Make predictions based upon data.

Materials needed:

. Cubes - ten per pupil

Comments and suggestions:

. The cubes will help pupils build the smaller models. For the large ones several pupils could work together to share cubes and strategi for solutions.

. Encourage the search for patterns, especially to determine the surf areas for the models in stage 10. Pupils may see many. As long as pattern is valid and does work, accept it as correct even though it vary. A possible pattern is given with each answer below.

Answers:

	1	2	3	4	5	10
A	6	12	18	24	30	60
B	6	10	14	18	22	42
C	6	24	54	96	150	600
D	6	14	24	36	50	150
E	6	18	34	54	78	258

Patterns:

A - number of cubes times 6.

B - add 4 each time after the first stage.

C - number of cubes on a side times itself times 6.

D -
$$\begin{array}{ccccccc} 6 & 14 & 24 & 36 & 50 & \ldots & 150 \\ & 8 & 10 & 12 & 14 & \ldots & 24 \end{array}$$

E - Side views will each be the stage number. Top and bottom will each be the odd number corresponding to the stage number, e.g. the 10th odd number.

Front and back views will each be the sum of consecutive odd numbers up to the stage number, e.g. the sum of the first 10 odd numbers.

THE CANDY COMPANY

Clara Fy's candy company created caramel bars.
Bad Billy bit a bite from the bars.

   a. Find the number of caramels needed for each whole caramel bar.
   b. Find the surface area of each whole caramel bar.
   c. Guess the number of caramels Bad Billy has bitten off each bar.

Cubes may help you do this activity.

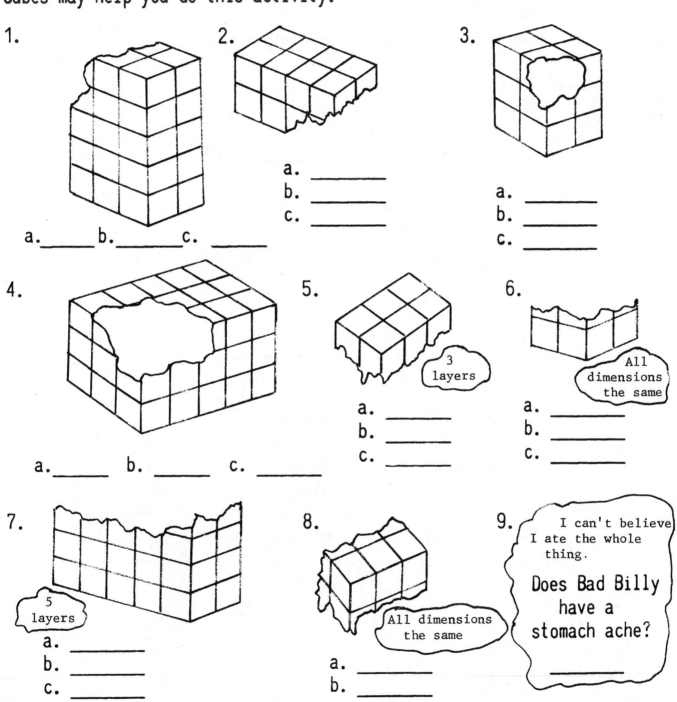

1. a._____ b._____ c._____

2. a._____
   b._____
   c._____

3. a._____
   b._____
   c._____

4. a._____ b._____ c._____

5. (3 layers)
   a._____
   b._____
   c._____

6. (All dimensions the same)
   a._____
   b._____
   c._____

7. (5 layers)
   a._____
   b._____
   c._____

8. (All dimensions the same)
   a._____
   b._____
   c._____

9. I can't believe I ate the whole thing.

Does Bad Billy have a stomach ache?

_____

PSM 81

The Candy Company

Mathematics teaching objectives:

. Find volume of rectangular boxes.

. Find surface area of rectangular boxes.

Problem-solving skills pupils might use:

. Use a drawing.

. Look for patterns.

. Search printed matter for needed information.

Materials needed:

. Cubes (optional)

Comments and suggestions:

. Pupils may use cubes to make each model. Some may draw in the lines needed to complete each model.

. The answers for part c are all estimates. In several views it isn't really known whether caramels hidden from view have been eaten or not.

Answers:

All part c answers are estimates. Pupil answers may vary.

1.	a.	30 caramels	b.	62 units	c.	4 caramels
2.	a.	16 caramels	b.	40 units	c.	4 caramels
3.	a.	12 caramels	b.	32 units	c.	1 caramels
4.	a.	45 caramels	b.	78 units	c.	6 caramels
5.	a.	18 caramels	b.	42 units	c.	12 caramels
6.	a.	8 caramels	b.	24 units	c.	4 caramels
7.	a.	50 caramels	b.	90 units	c.	20 caramels
8.	a.	27 caramels	b.	54 units	c.	21 caramels
9.	Probably					

Grade 6

VI. DECIMALS

VI. DECIMALS

This section on decimals follows the development started in the fifth-grade materials. Those activities involved naming, ordering, comparing, equating, adding, and subtracting decimals. Several models were used: grids, number lines, and money. No formal algorithms were given - only hands-on or discovery activities that lead toward the algorithms. Many of the activities that follow can be finished quickly and will need supplementing from textbooks or other sources - as long as the material is consistent with the model used.

These materials contain five readiness activities on naming and ordering decimals and activities stressing estimation in computation problems. The next four activities use a grid model to show how the dimensions and area of rectangles on the grid relate to multiplication and division facts. The intent is to give pupils the feel for multiplying "tenths times tenths and getting hundredths" as an answer. Finally four activities use a calculator to help pupils discover the "rules" for placing and moving decimal points in multiplication and division problems.

The development used is a decimals-only approach. No attempt is made to relate decimals and fractions. Several reasons for this approach seem valid:

1. The calculator encourages use of decimals.

2. The movement to convert to the metric system suggests more emphasis on decimals than fractions.

3. Pupils may relate better to decimals as an extension of the whole-number system rather than conversions from and to fractions.

Several other decimal activities are included in the <u>Drill</u> <u>And</u> <u>Practice</u> section. These might best be used after pupils have had and understand the concepts presented in both the fifth and sixth-grade <u>Decimals</u> units.

The average American
family has
4.4 members.

HUNDREDTHS BOARD

1. Rene changed her Hundreds Board by labeling it with decimals. Fill in the missing parts for her.

2. Rene invented a puzzle. The pattern goes like this:

$.38 \longrightarrow = .39$

$.52 \uparrow = .42$

$.16 \searrow = .27$

$.97 \nwarrow = .86$

What does each arrow mean?

.01	.02	.03							
.11			.14	.15	.16				
	.22					.27	.28	.29	
		.33		.35					.40
	.42		.44		.46		.48		
				.55					
	.62				.66				.70
				.75		.77			
	.82						.88		
				.95				.99	1.00

3. Solve these puzzles Rene made up.

   a. $.82 \rightarrow$ = ___   b. $.56 \nearrow$ = ___   c. $.40 \leftarrow$ = ___

   d. $.95 \nwarrow$ = ___   e. $.01 \searrow$ = ___   f. $.19 \downarrow$ = ___

4. Solve these more difficult problems of Rene's.   a $.38 \leftarrow \leftarrow \uparrow$ = ___

   b. $.67 \nwarrow \nearrow \nwarrow$ = ___   c. $.17 \downarrow \downarrow \searrow$ = ___   d. $.33 \nearrow \searrow \swarrow \nwarrow$ = ___

   e. $.99 \nwarrow \nwarrow \nwarrow$ = ___   f. $.70 \uparrow \leftarrow \downarrow \rightarrow$ = ___   g. $.24 \downarrow \rightarrow \downarrow \rightarrow$ = ___

5. Rene said each of these puzzles can be expressed with fewer symbols. Rewrite each one.

   a. $.05 \downarrow \searrow \rightarrow \uparrow$ = ___ ,   .05 _____

   b. $.63 \nwarrow \downarrow \uparrow \leftarrow \searrow$ = ___ ,   .63 _____

   c. $.91 \rightarrow \uparrow \uparrow \rightarrow \uparrow \leftarrow \leftarrow$ = ___ ,   .91 _____

6. Create your own puzzles for these decimals.

   a. .14 _____ = .33   b. .72 _____ = .65

7. Create your own puzzles using the <u>fewest</u> number of symbols.

   a. .43 _____ = .76   b. .67 _____ = .34   c. .41 _____ = .40

Hundredths Board

Mathematics teaching objectives:

. Write decimals up to 1.00 .

. Order decimals.

Problem-solving skills pupils <u>might</u> use:

. Search printed materials for needed information.

. Look for patterns and/or properties.

. Satisfy one condition at a time.

Materials needed:

. None

Comments and suggestions:

. The activity shows how decimals can be directly related to whole numbers.

. After completion of the activity, have pupils "define" what an $\uparrow$, etc. means. An answer like subtract one-tenth is sufficient.

Answers:

2. $\rightarrow$ means add .01, $\uparrow$ means subtract .1,

    $\searrow$ means add .11, $\nwarrow$ means subtract .11

3. a. .83   b. .47   c. .39   d. .84   e. .12   f. .29

4. a. .26   b. .36   c. .48   d. .33   e. .66   f. .70   g. .46

5. a. .05 $\searrow$ $\rightarrow$ = .17   b. .63 $\leftarrow$ = .62   c. .91 $\uparrow\uparrow\uparrow$ = .61

6. Answers may vary. Possible solutions are shown below.

    a. .14 $\downarrow\downarrow$ $\leftarrow$ = .33   b. .72 $\nearrow$ $\searrow$ $\nearrow$ = .65

7. a. .43 $\searrow\searrow\searrow$ = .76   b. .67 $\nwarrow\nwarrow\nwarrow$ = .34

    c. .41 $\nearrow \rightarrow \rightarrow \rightarrow \rightarrow \rightarrow \rightarrow \rightarrow$ = .40

ORDERING DECIMALS

You will get the point of this activity
if you follow directions.

1. Put a decimal point in each middle number so the three numbers
   are in order, smallest to largest. You may need extra zeros.

   a. 1, 314, 10   d. .1, 572, 1

   b. 10, 6275, 100   e. .01, 00384, .1

   c. 100, 1098, 1000   f. .001, 428, .01

2. Put a decimal point in each middle number so the three numbers
   are in order, smallest to largest. You may need extra zeros.

   a. .5, 538, .6   e. .03, 037, .04

   b. .99, 998, 1.00   f. .44, 44, 44

   c. 0.0, 7, .1   g. 1.00, 1001, 1.01

   d. .9, 987, 1.0   h. .003, 38, .004

3. For each part of problem 2, circle the number that the middle
   number is closer to after you have placed the decimal point.

4. Put one digit in each blank so the three numbers are in order
   from smallest to largest.

   a. .4, .__ __, .5   d. 9.0, 9. __ 7, 9.1

   b. 2.01, 2.0 __ __, 2.02   e. 0.0, 0. __ __, 0.1

   c. .50, .5 __ __, .51   f. .99, __.__, 1.1

PSM 81

-175-

Ordering Decimals

Mathematics teaching objectives:

. Order decimals by placing a decimal point.

. Compare decimals.

Problem-solving skills pupils _might_ use:

. Make decisions based upon data.

. Reason from what you already know.

. Find another answer when more than one is possible.

Materials needed:

. None

Comments and suggestions:

. Textbook exercises often give several decimals to be ordered from smallest to largest. If the decimals have the same number of decimal places, pupils will ignore the decimal and arrange the numbers as if they were whole numbers - which isn't bad but little understanding of decimals is promoted. By creating their own decimals, pupils gain more understanding.

. Several problems in (4) have multiple answers. Encourage pupils to list all possible answers.

Answers:

1. a. 3.14
   b. 62.75
   c. 109.8
   d. .572
   e. 0.0384
   f. .00428

2. a. .538
   b. .998
   c. .07
   d. .987
   e. .037
   f. 4.4
   g. 1.001
   h. .0038

3. a. .5
   b. 1.00
   c. .1
   d. 1.0
   e. .04
   f. .44
   g. 1.00
   h. .004

4. a. any of .41, ..., .49
   b. any of 2.011, ..., 2.019
   c. any of .501, ..., .509
   d. only 9.07
   e. any of 0.01, ..., 0.09
   f. only 1.0

ABOUT RIGHT

For each problem, circle the most reasonable answer A, B, or C.
No correct answers are given, so you shouldn't work the problems.

	A	B	C
1. 3.8 + 3.9	.8	80	8
2. .7 + .9	0	1	2
3. 2.1 + 5.2	.7	7	10
4. 6.3 + .1 + .24	6	85	9
5. 8.1 + 4.8	.2	13	120
6. 3.9 + 4.03	8	4	41
7. 8.1 + 5.73	65	6	14
8. 6.21 + 4.18	10	100	1
9. 9.84 + 7.9 + .98	19	12	28
10. .12 + .13 + .14	39	0	3.9

<u>About</u> <u>Right</u>

Mathematics teaching objectives:
. Estimate.
. Mentally add decimals and/or whole numbers.

Problem-solving skills pupils <u>might</u> use:
. Make reasonable estimates.

Materials needed:
. None

Comments and suggestions:
. Many pupils would rather compute exact answers than have to
  estimate. The fear of being wrong inhibits them. A positive
  class atmosphere and a series of activities like this one and
  the two titled "Be Reasonable" and "It's Missing" might help.
  In each pupils get practice in rounding up or rounding down to
  convenient amounts and then using those amounts to get an
  estimate.

Answers:
    1. C-8
    2. C-2
    3. B-7
    4. A-6
    5. B-13
    6. A-8
    7. C-14
    8. A-10
    9. A-19
  10. A-0 This answer should promote some discussion. Pupils may
             not think 0 is a number that can be used as an estimate.

BE REASONABLE

1. Circle the most reasonable answer for each problem. No
   correct answers are given so do not try to work the problems out.

		a	b	c
a.	38 x .8	300	40	3.8
b.	12.2 ÷ 4.1	50	30	3
c.	2.9 x 5.2	16	1500	8.1
d.	19.76 ÷ .43	0	15	40
e.	83.9 x 1.03	9	85	839
f.	57 ÷ .89	5.8	58	580
g.	.76 x 9.2	8	63	10
h.	4.6 ÷ 9.2	20	.4	36
i.	9.27 x .52	50	5	.5
j.	4.23 ÷ 2.4	200	20	2.0

2. Fill in each blank with a <u>decimal</u> that will give the reasonable
   answer shown. ≈ means "is about equal to."

   a. _____ x 5.2 ≈ 30   d. 78 ÷ _____ ≈ 80

   b. _____ ÷ 1.1 ≈ 8.9   e. _____ x .96 ≈ 50

   c. 36.9 x _____ ≈ 20   f. 100 ÷ _____ ≈ 26

3. Each reasonable answer below has two choices for the problem.
   Circle the problem that best gives the reasonable answer.

		a	b			a	b
a.	50	50 x .89	40 x .89	d.	36	39 ÷ .86	39 ÷ 1.04
b.	60	100 ÷ 1.2	80 ÷ 1.2	e.	25	5.1 x 5.4	5.1 x 4.9
c.	30	42.3 x 1.14	42.3 x .75	f.	11	90 ÷ 10.2	90 ÷ 9.8

<u>Be Reasonable</u>

Mathematics teaching objectives:

   . Estimate.

   . Mentally compute with decimals and whole numbers.

Problem-solving skills pupils <u>might</u> use:

   . Make reasonable estimates.

Materials needed:

   . None

Comments and suggestions:

   . Many pupils would rather compute exact answers than have to estimate. The fear of being wrong inhibits them. A positive class atmosphere and a series of activities like this one and the two titled "About Right" and "It's Missing" might help. In each pupils get practice in rounding up and rounding down to convenient amounts and then using those amounts to get an estimate.

   . Pupils need to acquire the feeling that it is o.k. to estimate. No estimate is ever wrong. Some estimates are just better than others.

   . In number (3), parts c and d emphasize the following:

        When multiplying by a number
           greater than 1, the product gets larger. .
           equal to 1, the product stays the same.
           less than 1, the product gets smaller.

        When dividing by a number
           greater than 1, the quotient gets smaller. .
           equal to 1, the quotient stays the same.
           less than 1, the quotient gets larger.

Answers:

1. a. b-40   d. c-40   g. a-8   j. c-2.0

     b. c-3   e. b-85   h. b-.4

     c. a-16   f. b-58   i. b-5

2. Answers will vary but should be close to:

     a. 6   b. 9   c. .75   d. .95   e. 50   f. 4

3. a. 50 x .89   c. 42.3 x .75   e. 5.1 x 4.9

     b. 80 ÷ 1.2   d. 39 ÷ 1.04   f. 90 ÷ 9.8

IT'S MISSING

1. For each of these problems only the first two digits of the
    answer are given. Also, the decimal point is missing.
    Estimate the answer and put the decimal point where it should be.

    a. 2.36 x 4.18 = 98   f. 1.56 ÷ 1.2 = 13

    b. 26.38 x 3.1 = 81   g. 39.36 ÷ 2.14 = 18

    c. 4.83 x 19 = 91   h. 15.2 ÷ 5.3 = 28

    d. .98 x .96 = 94   i. 1.283 ÷ .89 = 14

    e. 1.87 x .48 = 89   j. .68 ÷ .79 = 86

2. For each of these problems the decimal point <u>may</u> be in the
    wrong place. Estimate each answer. If the decimal point is
    correct, mark C . If it is not correct, put it in the
    correct place.

    a. 5.29 x 3.6 = 190.44   f. 1.69 ÷ 1.3 = .13

    b. 15.79 x 5.1 = 80.529   g. 47.712 ÷ 6.72 = .71

    c. 6.48 x 23 = 14.904   h. 18.4 ÷ 9.2 = .2

    d. 87 x .93 = 8.091   i. 3.325 ÷ .95 = 35

    e. 56 x .52 = .2912   j. .4816 ÷ .43 = .112

3. Invent a problem with an answer close to the given problem.
    Fill each blank with a single digit. Place the decimal points
    properly.

    a. ___ ___ x ___ ___ = 15.81

    b. ___ ___ ___ x ___ ___ = 14.875

    c. ___ ___ ___ ___ ÷ ___ ___ = 2.1

    d. ___ ___ ___ ÷ ___ = 1.8

It's Missing

Mathematics teaching objectives:

. Estimate.

. Mentally multiply and divide decimals and whole numbers.

Problem-solving skills pupils might use:

. Make reasonable estimates.

. Invent problems which can be solved by certain solution procedures.

Materials needed:

. None

Comments and suggestions:

. Many pupils would rather compute exact answers than have to estimate.
The fear of being wrong inhibits them. A positive class atmosphere
and a series of activities like this one and the two titled "About
Right" and "Be Reasonable" might help. In each pupils get practice
in rounding up and rounding down to convenient amounts and then using
those amounts to get an estimate.

. This activity gives partial and/or incorrect answers. Pupils have
to focus on the estimation process to correctly place the decimal
point.

Answers:

1.	a.	9.8		f.	1.3
	b.	81.		g.	18.
	c.	91.		h.	2.8
	d.	.94		i.	1.4
	e.	.89		j.	.86

2.	a.	19.044		d.	80.91		g.	7.1
	b.	C		e.	29.12		h.	2.
	c.	149.04		f.	1.3		i.	3.5
							j.	1.12

3. Answers may vary. Pupils could exchange
papers to check these problems.

DECIMAL PARTS 1

The large square represents 1 whole.

1. Use the vertical marks to divide the square into strips. Each strip represents what decimal part of the square?

   _____

2. Use the horizontal marks to divide the square into smaller squares. Each small square represents what decimal part of the large square?

   _____

3. For each rectangle shown, write the vertical dimension (in tenths), the horizontal dimension (in tenths), and the shaded part of the large square (in hundredths).

   a. [pattern] .3 , .4 , .12

   b. [pattern] ___ , ___ , ___

   c. [pattern] ___ , ___ , ___

   d. [pattern] ___ , ___ , ___

   e. [pattern] ___ , ___ , ___

   f. [pattern] ___ , ___ , ___

   g. [pattern] ___ , ___ , ___

   h. the whole square   ___ , ___ , ___

-183-

Decimal Parts 1

Mathematics teaching objectives:

. Indicate tenths and hundredths using a grid model.

. Relate dimensions (tenths) of a rectangle to its area (hundredths).

Problem-solving skills pupils might use:

. Make and use a drawing.

. Make predictions based upon data.

. Find another answer when more than one is possible.

Materials needed:

. None

Comments and suggestions:

. This activity relates the dimensions of a rectangle to the number of small squares contained in the rectangle. Without saying so, this is an application of the multiplication of decimals. In the lessons in this sequence pupils discover that "tenths multiplied by tenths" yields an answer in hundredths. Likewise "hundredths divided by tenths" yields an answer in tenths.

Answers:

1. one-tenth

2. one-hundredth

3. a. .3, .4, .12   d. .3, .6, .18   g. .1, .4, .04
   b. .2, .6, .12   e. .4, .4, .16   h. 1.0, 1.0, 1.00
   c. .5, .2, .10   f. .7, .4, .28

4. Answers will vary. Problems a-d shade .80 of the square.
   e. .2 or .20

5. a. .56   d. .54
   b. .36   e. .64
   c. .24   f. .21

6. a. .49   d.-f. Answers will vary.
   b. .7   g. .8
   c. .5

Decimal Parts 1 (cont.)

4. Shade rectangles with these dimensions. What decimal part of the large square is shaded in each rectangle? No overlapping is allowed.

    a. .2 high, .4 long, ____

    b. .5 high, .3 long, ____

    c. .8 high, .4 long, ____

    d. .5 high, .5 long, ____

    e. What decimal part of the large square is not shaded? _____

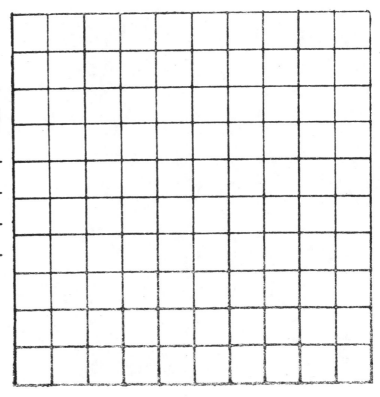

5. Predict the decimal part of the large square these rectangles would shade. Use the above square to check your predictions.

    a. .7 by .8 ____   d. .9 by .6 _____

    b. .4 by .9 _____   e. .8 by .8 _____

    c. .6 by .4 _____   f. .7 by .3 _____

6. Predict the missing dimension. Use the above square to check.

	a	b	c	d	e	f	g
Height	.7	.4	___	___	___	___	___
Length	.7	___	.9	___	___	___	1.0
Decimal part	___	.28	.45	.48	.36	.24	.80

DECIMAL PARTS 2

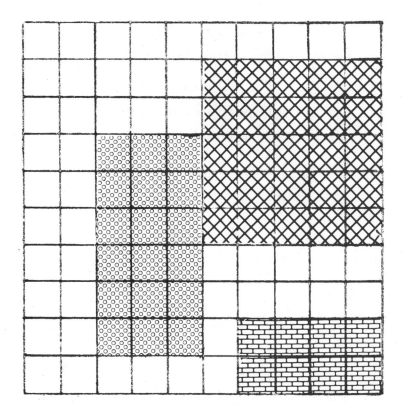

In the rectangle shaded 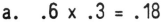, these facts can be seen:

a. .6 x .3 = .18
b. .3 x .6 = .18
c. .18 ÷ .3 = .6
d. .18 ÷ .6 = .3

Write multiplication and division facts for

a.  _____ , _____

b.  _____ , _____ ,

_____ , _____

2. Write one multiplication and one division fact for each of these rectangles.

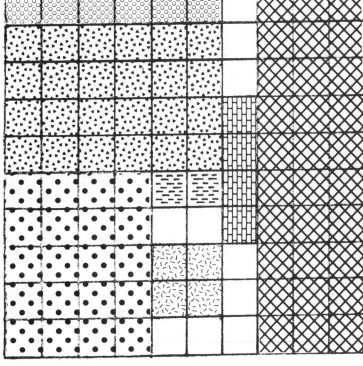

a.  _____ , _____

b. ▦ _____ , _____

c. ▨ _____ , _____

d. ⊡ _____ , _____

e. ⊠ _____ , _____

f. ⊞ _____ , _____

g. ▨ _____ , _____

h. If the unshaded squares were re-arranged into one rectangle, what rectangles could be formed? _____

Decimal Parts 2

Mathematics teaching objectives:

. Relate multiplication and division facts to dimensions of a
rectangle.

Problem-solving skills pupils might use:

. Use a drawing.

Materials needed:

. None

Comments and suggestions:

. This activity relates the dimensions of a rectangle to appropriate
multiplication and division facts. The lessons in this sequence
help pupils discover that "tenths multiplied by tenths" yields an
answer in hundredths and "hundredths divided by tenths" yields
an answer in tenths.

Answers:

1. a. $.5 \times .5 = .25$, $.25 \div .5 = .5$

    b. $.2 \times .4 = .08$, $.4 \times .2 = .08$, $.08 \div .2 = .4$, $.08 \div .4 = .2$

2. a. $.1 \times .6 = .06$, $.6 \times .1 = .06$, $.06 \div .1 = .6$, $.06 \div .6 = .1$

    b. $.1 \times .2 = .02$, $.2 \times .1 = .02$, $.02 \div .1 = .2$, $.02 \div .2 = .1$

    c. $.4 \times .6 = .24$, $.6 \times .4 = .24$, $.24 \div .4 = .6$, $.24 \div .6 = .4$

    d. $.5 \times .4 = .20$, $.4 \times .5 = .20$, $.20 \div .5 = .4$, $.20 \div .4 = .5$

    e. $1.0 \times .3 = .30$, $.3 \times 1.0 = .30$, $.30 \div 1.0 = .3$, $.30 \div .3 = 1.0$

    f. $.4 \times .1 = .04$, $.1 \times .4 = .04$, $.04 \div .1 = .4$, $.04 \div .4 = .1$

    g. $.2 \times .2 = .04$, $.04 \div .2 = .2$

    h. $.2$ by $.5$, $.5$ by $.2$, $.1$ by 1.0, 1.0 by $.1$

DECIMAL PARTS 3

1. Four medium squares are placed together to make a large square. Also, each medium square is divided into small squares.

    a. Each small square is what decimal part of a medium square?

    _____

    b. How many hundredths are shown in a medium square?

    _____

    c. How many hundredths are shown in the large square?

    _____

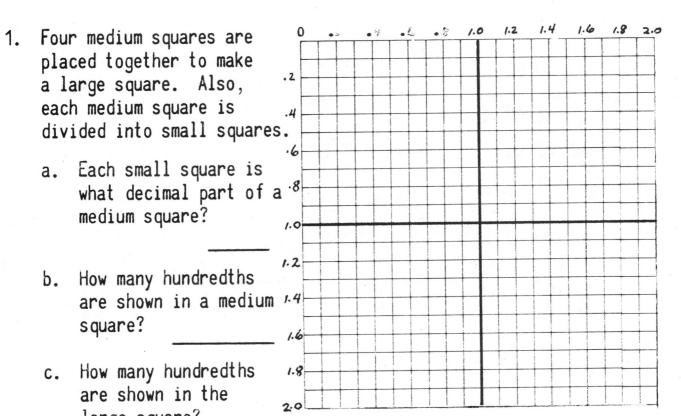

2. For each rectangle shown, write the vertical dimension, the horizontal dimension, and the shaded amount.

    a. ▦   ___ , ___ , ___

    b. ▧   ___ , ___ , ___

    c. ▨   ___ , ___ , ___

    d. ▧   ___ , ___ , ___

    e. ▨   ___ , ___ , ___

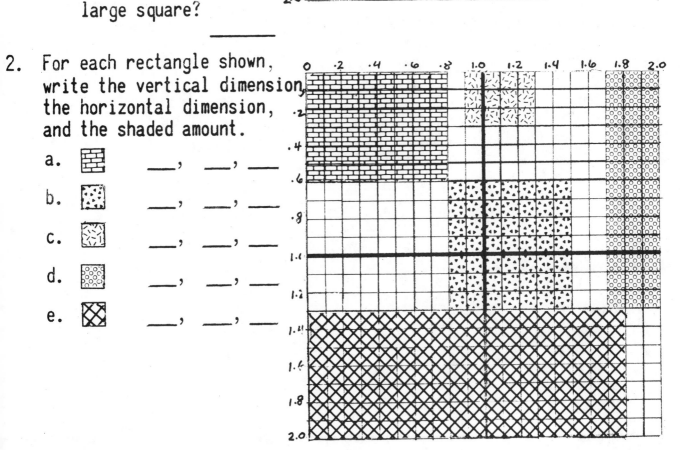

-189-

Decimal Parts 3

Mathematics teaching objectives:

. Relate dimensions (tenths) of a rectangle to its area (hundredths).

Problem-solving skills pupils might use:

. Use a drawing.

Materials needed:

. None

Comments and suggestions:

. This activity relates dimensions of a rectangle to the number of small squares contained in the rectangle. Four squares are placed together to get a large grid with dimensions of 2.0 by 2.0 and an area of 4.00. With this grid rectangles with dimensions of greater than 1.0 are possible.

Answers:

1. a. one-hundredth or .01
    b. one hundred-hundredths or 1.00
    c. four hundred-hundredths or 4.00

2. a. .6, .8, .48
    b. .7, .7, .49
    c. .3, .4, .12
    d. 1.3, .3, .39
    e. .7, 1.8, 1.26

DECIMAL PARTS 4

1. Write one multiplication
   and one division fact
   for each rectangle.

   a. ⊞ _____ , _____

   b. ▨ _____ , _____

   c. ⊠ _____ , _____

   d. ▨ _____ , _____

   e. ▨ _____ , _____

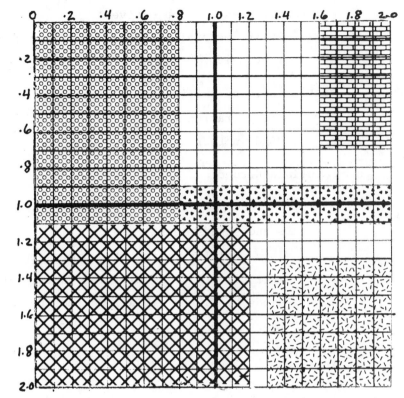

2. Shade rectangles to show
   these facts. Complete
   each statement. Over-
   lapping is allowed.

   a. .8 x .4 = _____

   b. .42 ÷ .6 = _____

   c. 1.08 ÷ .9 = _____

   d. .5 x .3 = _____

   e. .63 ÷ .7 = _____

   f. .2 x .9 = _____

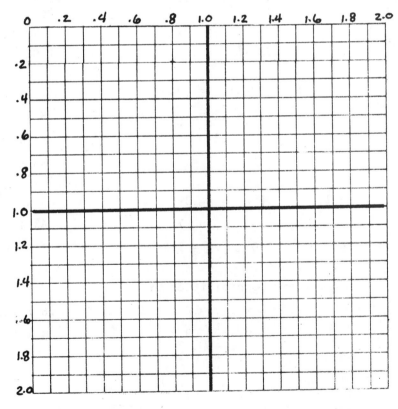

3. Write two different
   multiplication facts for
   each of these.

   a. .24 _____ , _____   b. .36 _____ , _____

   c. .48 _____ , _____   d. .88 _____ , _____

-191-

Decimal Parts 4

Mathematics teaching objectives:

. Relate multiplication and division facts to dimensions of a
  rectangle.

Problem-solving skills pupils might use:

. Use a drawing.

. Find another answer when more than one is possible.

Materials needed:

. None

Comments and suggestions:

. This activity relates the dimensions of a rectangle to appropriate
  multiplication and division facts. The expanded grid with dimensions of 2.0 by 2.0 allows rectangles to have dimensions greater
  than 1.0.

Answers:

1. a. .7 x .4 = .28,   c. .9 x 1.2 = 1.08
        .4 x .7 = .28,   1.2 x .9 = 1.08
        .28 ÷ .7 = .4,   1.08 ÷ .9 = 1.2
        .28 ÷ .4 = .7   1.08 ÷ 1.2 = .9

    b. .2 x 1.2 = .24,   d. 1.1 x .8 = .88   e. .7 x .7 = .49
        1.2 x .2 = .24   .8 x 1.1 = .88   .49 ÷ .7 = .7
        .24 ÷ .2 = 1.2   .88 ÷ 1.1 = .8
        .24 ÷ 1.2 = .2   .88 ÷ .8 = 1.1

2. a. .8 x .4 = .32   d. .5 x .3 = .15
    b. .42 ÷ .6 = .7   e. .63 ÷ .7 = .9
    c. 1.08 ÷ .9 = 1.2   f. .2 x .9 = .18

3. Some possibilities are:
    a. .4 x .6,   .2 x 1.2   c. .6 x .8,   .3 x 1.6
    b. .6 x .6,   .9 x .4   d. .8 x 1.1,   .4 x 2.2

A CALCULATOR DISCOVERY

Use a calculator to do these multiplications. Discover how the
calculator places the decimal point.

Set 1 - a. 3.5 x 9 = _____   b. 4.8 x 6 = _____

     c. 14.3 x 7 = _____   d. 23.8 x 19 = _____

Set 2 - a. 3.7 x 5.1 = _____   b. 12.9 x 4.6 = _____

     c. 26.5 x 14.3 = _____   d. 13.3 x 13.3 = _____

Set 3 - a. 3.64 x 2.3 = _____   b. 5.19 x 8.2 = _____

     c. 12.17 x 9.7 = _____   d. 13.89 x 12.3 = _____

Set 4 - a. .37 x .52 = _____   b. 3.14 x .83 = _____

     c. 1.68 x 1.68 = _____   d. 12.35 x .25 = _____

Set 5 - a. .375 x 3 = _____   b. 5.813 x 9 = _____

     c. 12.111 x 8 = _____   d. 5.314 x 11 = _____

Study the problems and answers. How does the calculator place
the decimal point? _____

First predict the number of decimal places for each of these
answers. Then check your prediction on the calculator.

Set 6 - a. 4.5 x 1.9 = _____   b. 3.625 x 7 = _____

     c. 18.4 x 103 = _____   d. 9.75 x 1.3 = _____

A Calculator Discovery

Mathematics teaching objectives:

. Use a calculator.

. Discover a "rule" for placing the decimal point in a product.

Problem-solving skills pupils _might_ use:

. Look for properties.

. Make explanations based upon data.

Materials needed:

. Calculator.

Comments and suggestions:

. By studying the problems and answers, pupils _may_ discover that
the number of decimal places in the product is the same as the
sum of the number of decimal places in the factors.

. If the discovery is not made, a teacher-led discussion is needed
to draw out the discovery.

Answers:

Set 1	a.	31.5	b.	28.8	c.	100.1	d.	452.2
Set 2	a.	18.87	b.	59.34	c.	378.95	d.	176.89
Set 3	a.	8.372	b.	42.558	c.	118.049	d.	170.847
Set 4	a.	.1924	b.	2.6062	c.	2.8224	d.	3.0875
Set 5	a.	1.125	b.	52.317	c.	96.888	d.	58.454

The calculator places the decimal point so the product has as many
decimal places as the factors.

Set 6	a.	2, 8.55		b.	3, 25.375
	c.	1, 1895.2		d.	3, 12.625

PATTERNS WITH A CALCULATOR 1

1. Use a calculator to do these problems.

   a. 314 x 10 = _____   e. 314 ÷ 1 = _____
   b. 314 x 100 = _____   f. 314 ÷ 10 = _____
   c. 314 x 1000 = _____   g. 314 ÷ 100 = _____
   d. 314 x 10,000 = _____   h. 314 ÷ 1000 = _____
   i. Predict the answer for these problems. Then check your
       predictions using the calculator.
   j. 314 x 100,000 = _____   k. 314 ÷ 10,000 = _____

2. Write yes if the answer is correct. Write no if the answer
   is incorrect. Check each with the calculator.

   a. 293 x 100 = 29,300 _____   e. 729 x 1000 = 72,900 _____
   b. 582 x 10 = 5820 _____   f. 338 ÷ 1000 = .338 _____
   c. 639 ÷ 100 = 63.9 _____   g. 915 ÷ 1 = 91.5 _____
   d. 138 ÷ 10 = 13.8 _____   h. 482 x 100 = 48200 _____

3. Predict the answer for each problem. Then check each with the
   calculator.

   a. 274 ÷ 100 = _____   e. 414 x 100 = _____
   b. 515 x 10 = _____   f. 888 ÷ 1 = _____
   c. 928 x 1000 = _____   g. 751 x 1000 = _____
   d. 378 ÷ 10 = _____   h. 128 ÷ 10 = _____

4. Here is the final exam for this lesson. First predict the
   answer. Then check with the calculator.

   a. 27 x 100 = _____   c. 5 ÷ 10 = _____
   b. 4529 ÷ 100 = _____   d. 3492 x 10 = _____

Patterns With A Calculator 1

Mathematics teaching objectives:

. Multiply and divide by powers of ten.

. Use a calculator.

Problem-solving skills pupils might use:

. Look for patterns.

. Make predictions based upon data.

Materials needed:

. Calculator

Comments and suggestions:

. Pupils use the calculator to discover patterns when multiplying and dividing by powers of ten.

Answers:

1. a.   3,140   e. 314
   b.   31,400   f. 31.4
   c. 314,000   g. 3.14
   d. 3,140,000   h. .314

   j. 31,400,000   k. .0314

2. a. Yes   e. No - 729,000
   b. Yes   f. Yes
   c. No - 6.39   g. No - 915
   d. Yes   h. Yes

3. a. 2.74   e. 41,400
   b. 5150   f. 888
   c. 928,000   g. 751,000
   d. 37.8   h. 12.8

4. a. 2700   c. .5
   b. 45.29   d. 34,920

PATTERNS WITH A CALCULATOR 2

1. Find answers for each set of problems.

Set 1	Set 2	Set 3
a. $12 \div 2 =$ ___	a. $288 \div 24 =$ ___	a. $28 \div 4 =$ ___
b. $24 \div 4 =$ ___	b. $144 \div 12 =$ ___	b. $280 \div 40 =$ ___
c. $48 \div 8 =$ ___	c. $72 \div 6 =$ ___	c. $2800 \div 400 =$ ___
d. $96 \div 16 =$ ___	d. $36 \div 3 =$ ___	d. $28000 \div 4000 =$ ___

What is true about all the answers to the problems in each set?

2. Use a calculator to find answers for each set of problems.

Set 1	Set 2
a. $4.554 \div .09 =$ ___	a. $187.5 \div 15 =$ ___
b. $45.54 \div .9 =$ ___	b. $18.75 \div 1.5 =$ ___
c. $455.4 \div 9 =$ ___	c. $1.875 \div .15 =$ ___

Set 3
a. $25.00 \div .0125 =$ ___
b. $2500 \div 1.25 =$ ___
c. $250,000 \div 125 =$ ___

Make up a problem that fits the pattern for each set.

Set 1 d. _____   Set 2 d. _____   Set 3 d. _____

3. Each problem in a set has the same answer. Circle the one you think would be the easiest to work using only paper and pencil.

Set 1	Set 2	Set 3
a. $355 \div 71$	a. $1.072 \div .08$	a. $262.5 \div 17.5$
b. $.355 \div .071$	b. $10.72 \div .8$	b. $26.25 \div 1.75$
c. $3.55 \div .71$	c. $1072 \div 80$	c. $2625 \div 175$

Find the correct answer for each set.

Set 1 _____   Set 2 _____   Set 3 _____

Patterns With A Calculator 2

Mathematics teaching objectives:

. Introduce equivalent problems.
. Discover a division property for dividing by a decimal.
. Use a calculator.

Problem-solving skills pupils _might_ use:

. Look for properties.
. Solve an easier but related problem.
. Invent problems which can be solved by certain solution procedures.

Materials needed:

. Calculator.

Comments and suggestions:

. The use or non-use of the calculator is important for this activity. It should be used in problem 2 and not used, except for checking, in problems 1 and 3.

. The property being discovered is that dividing by a whole number is easier than dividing by a decimal when no calculator is available. Eventually, multiplying by an appropriate power of 10 (or moving the decimal point) allows pupils to write the division problem in this form. The very important problem-solving skill used in this activity is solving an easier but related problem.

Answers:

1. Set 1   All answers are 6.

    Set 2   All answers are 12.

    Set 3   All answers are 7.

2. Set 1   All answers are 50.6 $.4554 \div .009$ Possible
    Set 2   All answers are 12.5 $1875 \div 150$ answers
    Set 3   All answers are 2000. $250 \div .125$ for part d.

3. Set 1 $355 \div 71 = 5$

    Set 2 $1072 \div 80 = 13.4$

    Set 3 $2625 \div 175 = 15$

In all these problems, some systematic adjustment is made to each part of the problem. A main purpose of the lesson is for pupils to examine the nature of the adjustment.

PATTERNS WITH A CALCULATOR 3

1. Look for patterns. Does each problem in a set have the same answer? Answer __yes__ or __no__. __DO NOT__ work the problems.

Set 1	Set 2	Set 3	Set 4
a. $1\overline{)2}$	a. $2.5 \div .5$	a. $170\overline{)6290}$	a. $380730 \div 2590$
b. $4\overline{)8}$	b. $25 \div 5$	b. $17\overline{)629}$	b. $3807.3 \div 25.9$
c. $16\overline{)32}$	c. $250 \div 50$	c. $1.7\overline{)62.9}$	c. $380.73 \div 2.59$
d. $64\overline{)128}$	d. $2500 \div 500$	d. $.017\overline{).629}$	d. $3.8073 \div .0259$

_____   _____   _____   _____

2. Circle the problem from each set which does not have the same answer as the other three. Look for patterns. __DO NOT__ work the problems.

Set 1	Set 2	Set 3	Set 4
a. $9 \div 3$	a. $.007\overline{)49}$	a. $8760 \div 120$	a. $1350\overline{)60750}$
b. $18 \div 6$	b. $.7\overline{)490}$	b. $876 \div 12$	b. $13.5\overline{)607.5}$
c. $36 \div 18$	c. $7\overline{)4900}$	c. $87.6 \div 1.2$	c. $1.35\overline{)6.075}$
d. $72 \div 24$	d. $70\overline{)49,000}$	d. $8.76 \div .012$	d. $.0135\overline{).6075}$

3. Does each problem in a set have the same answer? Which do you think will be the easiest to do?

Set 1	Set 2	Set 3
a. $1.56 \div .3$	a. $.008\overline{)1.720}$	a. $1.2\overline{)648}$
b. $.156 \div .03$	b. $8\overline{)1720}$	b. $.12\overline{)64.8}$
c. $15.6 \div 3$	c. $.8\overline{)172.0}$	c. $12\overline{)6480}$

Set 4		Set 5	
a. $1150 \div 25$		a. $.81\overline{).6399}$	
b. $1.150 \div .025$		b. $81\overline{)63.99}$	
c. $115 \div 2.5$		c. $.081\overline{).06399}$	

Patterns With A Calculator 3

Mathematics teaching objectives:

. Use equivalent problems

. Use a division property for dividing by a decimal.

. Use a calculator.

Problem-solving skills pupils _might_ use:

. Look for patterns.

. Solve an easier but related problem.

Materials needed:

. Calculator

Comments and suggestions:

. The previous page focused on problem sets with the same answers. This page asks pupils to decide by inspection, whether several problems do have the same answers. They need to look at the adjustments done to each part of the problem.

. Note that 1 and 2 do _not_ ask pupils to solve the problems.

. Problem 3 emphasizes the main purpose of the lesson - when doing a division by a decimal, make an equivalent problem having a whole number divisor. It will be easier to work.

Answers:

	Set 1	Set 2	Set 3	Set 4
1.	Yes	Yes	Yes	Yes
2.	$36 \div 18$	$.007 \overline{)\ 49}$	$8.76 \div .012$	$1.35 \overline{)\ 6.075}$
3.	$15.6 \div 3 = 5.2$	$8 \overline{)\ \overset{215}{1720}}$	$12 \overline{)\ \overset{540}{6480}}$	$1150 \div 25 = 46$

Set 5

$$81 \overline{)\ \overset{.79}{63.99}}$$

Grade 6

VII. PROBABILITY

Experimental probability provides an excellent opportunity to emphasize these problem-solving skills: collect data, make a table, use a model, make predictions and in general, guess and check. As pupils suggest ways of solving the problems (or as they describe how they solved the problems) their answers can be translated into standard problem-solving phrases. For example, "Keep track of what you get" can be verbalized by the teacher as "Make a table." Specific suggestions for emphasizing the problem-solving skills are given in the comments and suggestions for each page.

Working with experimental probability in grades 5 and 6 is also instructionally sound. Usually pupils are not ready for any formal probability like "The probability of rolling a sum of 7 is $\frac{7}{36}$," until somewhere between the seventh and tenths grades. Experimental probability provides some background for later work with formal probability and, more importantly, it provides experience with a method of solving problems that is very useful in the real world. To avoid preconceived notions, like "Heads are luckier than tails," the activities in this section use situations with which pupils are not familiar. Each activity allows pupils to collect enough data to predict the outcome.

<u>Using</u> <u>The</u> <u>Activities</u>

With enough equipment, the entire class can complete one activity and the results can be compared and combined. Another option is to have small groups rotate through activity stations which are all on experimental probability or are on a mixture of topics.

The extra materials needed for these activities include:
   . regular dice, two dice per pair of pupils.
   . 20 $1 bills from a Monopoly game per pair of pupils.
   . several coins
   . two spinners per pair of pupils. See "Spin That Spinner"
      for the master.

Pupils may need review on an efficient way to tally, ⅲⅲ groups by fives.

PAPER, SCISSORS, OR ROCKS

Needed: 3 players

Do you know the paper, scissors, or rocks game? All players make
a fist and together count to four. On the count of four, each
player shows either

   a. <u>paper</u> by showing four fingers

   b. <u>scissors</u> by showing two fingers

   c. <u>rock</u> by keeping the fist together.

1. Play this game with two partners. Decide who is

   Player A _____   Player B _____   Player C _____

2. Play the game 25 times with these rules:

   a. Player A gets a point if all players show the same sign.
   b. Player B gets a point if only two players show the same sign.
   c. Player C gets a point if all players show a different sign.

3. Tally the winning
   points in this
   table.

Player	Tally	Total
A		
B		
C		

4. Is this a fair game? _____   Which player would you rather be? _____

5. Make a list of the ways three players could show the signs.

6. What change in the points awarded will make the game more fair?

7. Play the game 25 times with your rules. Record the results.

<u>Paper</u>, <u>Scissors</u>, or <u>Rocks</u>

Mathematics teaching objectives:

. Develop probability concepts.
. Practice recording skills.

Problem-solving skills pupils <u>might</u> use:

. Collect data needed to solve the problem.
. Make and use a table.
. Make a systematic list.

Materials needed:

. None

Comments and suggestions:

. This lesson involves a lot of activity. Eight or more groups of pupils each counting, slapping their hands together and showing the signs can be quite noisy. Be sure to be in control of this activity from the beginning.

. Before starting, you can explain and model the activity with two pupils. The condition where player B scores a point (only two showing the same sign) is often misunderstood. Ask pupils if they might have a preference for being player A, B or C.

. Problem 5 asks for a list of all the different ways the three players could show the signs. This activity should probably be teacher directed so none of the 27 possibilities are missed.

. Combined class results usually show the decided advantage to player B.

Answers:

3. Answers will vary. One group got these results from playing the game 25 times:

Player	Tally	Total
A	II	2
B	IHT IHT IHT I	16
C	IHT II	7

4. The game is <u>not</u> fair. It's best to be Player B.

5.

A	B			C
PPP	PSS	SPS	SSP	PSR
SSS	PRR	RPR	RRP	PRS
RRR	SPP	PSP	PPS	SPR
	SRR	RSR	RRS	SRP
	RPP	PRP	PPR	RPS
	RSS	SRS	SSR	RSP

6. The game becomes "theoretically" fair if Player A receives 6 points, Player C receives 3 points and Player B receives 1 point for a win.

DOUBLES IN MONOPOLY

Needed: A banker and a player
20 $1 bills from a Monopoly game
Two dice

Rolling doubles is important in Monopoly.

- You get an extra turn.
- You get sent to jail for 3 doubles in a row.
- You can get out of jail by rolling doubles.

1. Do you think rolling doubles is easy? _____

2. Try this experiment.

   a. Give the player and the banker 10 $1 bills each.

   b. Have the player roll the dice.

      . If the dice show doubles, the banker pays the player
$3.

      . If the dice do not show doubles, the player pays the
banker $1.

   c. Repeat 20 times or until someone runs out of money.

   d. Record below.

Roll	1	2	3	4	5	6	7	8	9	10	11	12	13	14	15	16	17	18	19	20
Doubles (Yes or No)																				
Player's $1 bills																				
Banker's $1 bills																				

3. Is this a fair experiment for the player? _____

4. Repeat the experiment with the banker paying $5 for doubles.
Is this a fair experiment for the banker?

Doubles In Monopoly

Mathematics teaching objectives:

. Develop probability concepts.

. Practice recording skills.

Problem-solving skills pupils <u>might</u> use:

. Collect data needed to solve the problem.

. Make and use a table.

Materials needed:

. 20 $1 bills from a Monopoly game

. 2 regular dice

Comments and suggestions:

. The $1 bills from a Monopoly game are motivational for the activity but not essential. The record keeping in the table will show the exchange of bills as the activity progresses.

Answers:

1. Answers will vary. Many pupils think doubles are difficult to get since so many board games require doubles to do special things.

2. Answers will vary. One experiment of 20 rolls gave these results.

Roll	1	2	3	4	5	6	7	8	9	10	11	12	13	14	15	16	17	18	19	20
Doubles (Yes or No)	N	Y	N	N	Y	N	N	N	N	N	N	N	N	Y	N	N	N	N	N	N
Player's $1 bills	9	12	11	10	13	12	11	10	9	8	7	6	5	8	7	6	5	4	3	2
Banker's $1 bills	11	8	9	10	7	8	9	10	11	12	13	14	15	12	13	14	15	16	17	18

3. The experiment is <u>not</u> fair for the player. Combined class results should emphasize this.

4. This experiment is "theoretically" fair for both players. The chance of <u>not</u> rolling doubles is five times the chance of rolling doubles. But since experimental results (in a small number of trials) seldom exactly agree with theoretical probability, combined class results should show a variety of answers.

SPIN THAT COIN

Needed: A coin

When Sally flipped a coin 50 times she got about the same number of heads and tails. She wondered what would happen if she would spin the coin.

What do you think? Check one of these statements.

_____ Heads and tails will show about the same.

_____ More heads will show.

_____ More tails will show.

1. Use one finger to hold the coin straight up.

2. Snap the coin on one edge to make it spin.

3. When the coin stops, record either H or T for the side that shows.

4. Repeat 30 times.

5. The final totals are H ____ and T ____ . Is this what you expected to happen? _____

6. Repeat the activity with a different coin.

7. Repeat the activity with a Canadian penny.

Spin That Coin

Mathematics teaching objectives:

. Develop probability concepts.

. Practice recording skills.

Problem-solving skills pupils _might_ use:

. Collect data needed to solve the problem.

. Make predictions based upon data.

Materials needed:

. Several coins

Comments and suggestions:

. Pupils realize that flipping a coin usually gives about the same number of heads as tails. Spinning a coin presents a situation they may not know anything about or perhaps they think the results will be the same as flipping.

. If the floor is _not_ carpeted, have pupils spin the coins on the floor. This avoids the problem of having the coins fall off the table or desk.

Answers:

4. & 5. Answers will vary but most coins will show more tails. The spinning coin tends to fall with the heavier side down. Although there is little difference the head side is heavier.

6. Answers will vary.

7. The Canadian penny tends to fall with the head side up more often.

SPIN THAT SPINNER

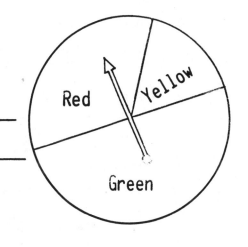

Needed: Two spinners like the one shown.

1. Use one spinner. Guess –

    a. which color will occur most often. ____

    b. which color will occur least often. ____

    c. Test your guesses by spinning
        24 times. Record your results.

Color	Tally	Total
Yellow		
Red		
Green		

2. Now use two spinners.

    a. If you spin both spinners, what color combinations could occur?

        _____ _____ _____ _____ _____ _____

    b. If you spin both spinners, guess which color combination
        will occur most often. ____

    c. Spin both spinners 48 times.
        Record your results.

    d. Is this what you expected?

        _____

Colors	Tally	Total
YY		
RR		
GG		
YR		
YG		
RG		

Spin That Spinner

Mathematics teaching objectives:

. Develop probability concepts.
. Practice recording skills.

Problem-solving skills pupils might use:

. Collect data needed to solve the problem.
. Make and use a table.

Materials needed:

. Two spinners as shown on the pupil page.

Comments and suggestions:

. If commercial spinners are not available, this homemade
spinner works reasonably well. Straighten out one
part of a paper clip. Place a pencil or pen in
the center of the spinner. Snap the straighten-
ed part with a finger.

. The results for the second experiment are usually
quite surprising. Most pupils predict that green-
green will occur most often, but it is usually
green-red. This happens because there is no distinction between
the two spinners as far as green-green is concerned. But the
green-red combination can be either green-red or red-green.

Answers:

1. c. Answers will vary. One
        experiment of 24 spins
        gave these results.

Color	Tally	Total
Yellow	卌	5
Red	卌 /	6
Green	卌 卌 ///	13

2. a. YY, RR, GG, YR, YG, RG

    b. Answers will vary. Most
        pupils will answer GG.

    c. Answers will vary. One
        experiment of 48 spins
        gave these results ⟶

    d. Probably not!

Colors	Tally	Total
YY	//	2
RR	///	3
GG	卌 卌 //	12
YR	卌 /	6
YG	卌 ///	8
RG	卌 卌 卌 //	17

-212-

Grade 6

VIII. CHALLENGES

The activities in the <u>Getting</u> <u>Started</u> section were very directed and
pupils were encouraged to use (although not completely restricted to) one
problem-solving skill at a time. The challenge problems in this section
leave the choice of the problem-solving method up to the pupil. The inten-
tion is to allow for and encourage individual differences, creativity, and
cooperation.

Let's look at how one challenge problem, "Addition Challenge" (page 229),
can be used in the classroom. As you read the example, notice how the teacher
does not structure or direct the methods pupils use, but that the teacher does
have these important functions:

. to help pupils <u>understand</u> the problem.
. to <u>listen</u> if pupils want to discuss their strategies.
. to <u>praise</u> and <u>encourage</u> pupils in their attempts, successful or not.
. to <u>facilitate</u> discussion of the problem and sharing of the strategies.
. to <u>give</u> hints or ask questions, if necessary.
. to <u>summarize</u> or <u>emphasize</u> methods of solution after pupils have
  solved the problem.

As we look in on Ms. Herman's classroom, the page "Addition Challenge"*
has just been distributed. It is near the end of the math period - about 10
minutes before recess. Pupils are used to seeing a different challenge each
week and know they will be given several days to work on the problem.

Ms. H: Here is the challenge problem for the week. I'll let you look at
         the puzzle, give you time to read the directions and get started.
         After three or four minutes, we'll discuss the problem to be sure
         we all understand it. (Waits) Who can explain how to complete
         the puzzle? Susan.

Susan: The puzzle has nine blank circles. I think a different digit goes
         in each circle.

Ms. H: Lee, you look puzzled.

Lee: Oh! I get it. The circles all have to add up to 999.

Ms. H: You and Susan explained the puzzle very well. Work on the first
         puzzle. See if you can solve it before recess.

Dale: (Hand raised)   Ms. Herman, I found one answer but I can't find
         any others.

Ms. H: It wouldn't be called a challenge if you knew right away, would it?
         Try to find some other solutions. If you get one, try to remember
         how you did it. This might help some of the rest of us. (Class is
         dismissed.)

*See page 229.

Later in the week. . .

Ms. H: How many of you found a solution to the challenge?   (Several
         pupils raise their hands.)   Scott?

Scott: I found three different answers.   May I write them on the board?
         (Writes the following.)

195	196	198
476	475	475
+ 328	+ 328	+ 326
999	999	999

Ms. H: Great, Scott.   I think as a class we need to decide if these are
         different answers.   What do you think?

Toby: I think those can be called the same because all Scott did was
         to change a couple of numbers around in one column.

Ms. H: Then you think a different solution needs different digits in
         the columns?

Toby: Yes.

Ms. H: Class, do you agree?   (Class nods yes.)   Would someone tell
         about the strategy used to get an answer?

Wanda: I just guessed and checked.

Terry: I wrote the numbers on slips of paper and moved them around
         until I got an answer.

Jim: I figured out that only the 1, 2 and 5 could be used in the
         left column but I must have been wrong because Scott's answer
         uses 1, 3 and 4.

Ms. H: You've all used good strategies.   Now you can work on the rest
         of the challenge.   Remember about what we said earlier this year.
         You should only spend about 15 minutes working on a single prob-
         lem if you aren't getting an answer or it doesn't look like you
         will get an answer.   At the end of 15 minutes, try to see if you
         can explain why an answer might not be possible.

Ms. Herman terminates this discussion at this point and introduces the class
to the geometry activities which she and the other sixth-grade teachers had
planned for the week.   She was planning later in the week to continue with
the discussion of the challenge.

The above approach to challenge problems also gives opportunities for practicing the following problem-solving skills:

. State the problem in your own words.

. Clarify the problem through careful reading and by asking questions.

. Share data and the results with other interested persons.

. Listen to persons who have relevant knowledge and experiences to share.

. Study the solution process.

. Invent new problems by varying an old one.

> # THE TEACHER MUST BE AN ACTIVE, ENTHUSIASTIC
> # SUPPORTER OF PROBLEM SOLVING

Using The Activities

Fifteen varied challenge problems are provided. Some teachers give them as a "challenge of the week" or as a Friday activity. On the day the challenge is given out, time should be spent on getting acquainted with the problem. On following days, a few minutes can be devoted to pupil progress reports. If there is little sign of progress, you can provide some direction by asking a key question or suggesting a different strategy. At appropriate times, the activity can be summarized by a class discussion of strategies used and some problem extensions.

One Plan For Using A Challenge (over a period of 1 or 2 weeks)

First day -

. Give out the challenge. (Possibly near the end of the period.)

. Let pupils read written directions and possibly discuss with a classmate.

. Clarify any vocabulary which seems to be causing difficulty. Ask a few probing questions to see if they have enough understanding to get started.

. Remind them that during a later math class, time will be used to look at the problem again.

Later in the week -

. Have pupils share their ideas.

. Identify the problem-solving skills suggested by these ideas.

. Conduct a brainstorming session if pupils do not seem to know how to get started.

. Suggest alternative strategies they might try.

. Give an extension to those pupils who have completed the challenge.

On a subsequent day -

. Allow some class time for individuals (or small groups) to work on the challenge. Observe and encourage pupils in their attempts.

. Try a strategy along with the pupils (if pupils seem to have given up).

Last day -

. Conduct a session where pupils can present the unsuccessful as well as the successful strategies they used.

. Possibly practice a problem-solving skill that is giving pupils difficulty; e.g., recording attempts, making a systematic list, or checking solutions.

Key problem-solving strategies that pupils have used in solving the problems are given in the comments for each problem. Your pupils might have additional ways of solving the problems.

A challenge problem for the teacher: Keep the quick problem solver from telling answers to classmates.

HEXAGON PUZZLE

Place the numbers 1-19 in the circles so the sum of
the rows with three circles around the outside is 23 and
the sum of the rows with five circles through the middle is 40.

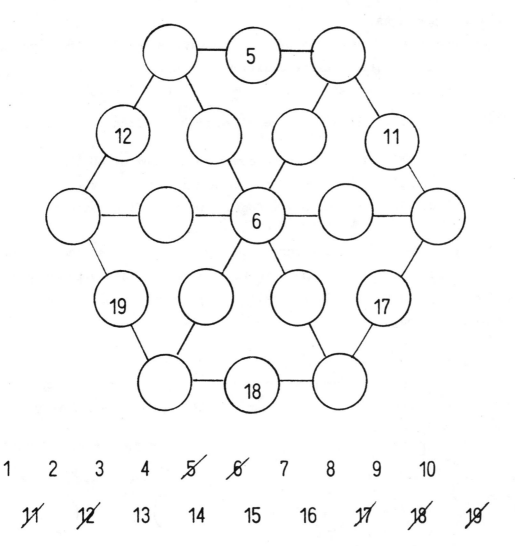

1   2   3   4   5̸   6̸   7   8   9   10

11̸   12̸   13   14   15   16   17̸   18̸   19̸

Hexagon Puzzle

Problem-solving skills pupils <u>might</u> use:
- Guess and check.
- Break a problem into manageable parts.
- Eliminate possibilities.

Comments and suggestions:
- Some students will break the problem into parts, finding the numbers in the outer ring first using addition and subtraction. After the outer ring is solved, six numbers remain. Pupils may follow these steps:
  - a. Find the sum of the given numbers already in a 5-circle row.
  - b. Find two numbers to complete the sum of 40.
  - c. Place the numbers and repeat for the other rows.
  - d. If one or more sum doesn't work, refine the choice of the numbers in part (b).

- Other pupils may do nothing more than guess and check at the placement of numbers. Markers with the numbers 1-19 may be used with guessing and checking to try possibilities. This procedure avoids a lot of erasing of wrong guesses.

Answers:
- Two distinct answers are possible. Of course, some numbers could also be switched across the 6. For example, the 15 and 14 could be switched in the first answer.

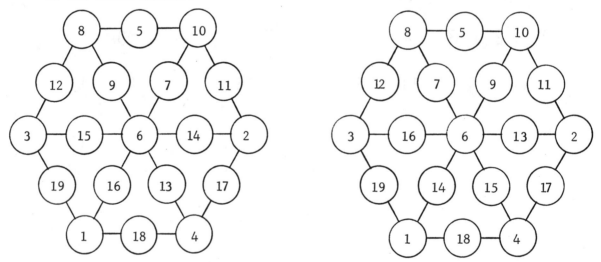

TRIANGLES AND RECTANGLES

Get some triangles like the one shown. For each problem below, you will need a triangle.

Get some scissors.

1. Make 1 cut to get 2 pieces that exactly cover the rectangle.

2. Make 2 cuts to get 3 pieces that exactly cover the rectangle.

3. Make 3 cuts to get 4 pieces that exactly cover the rectangle.

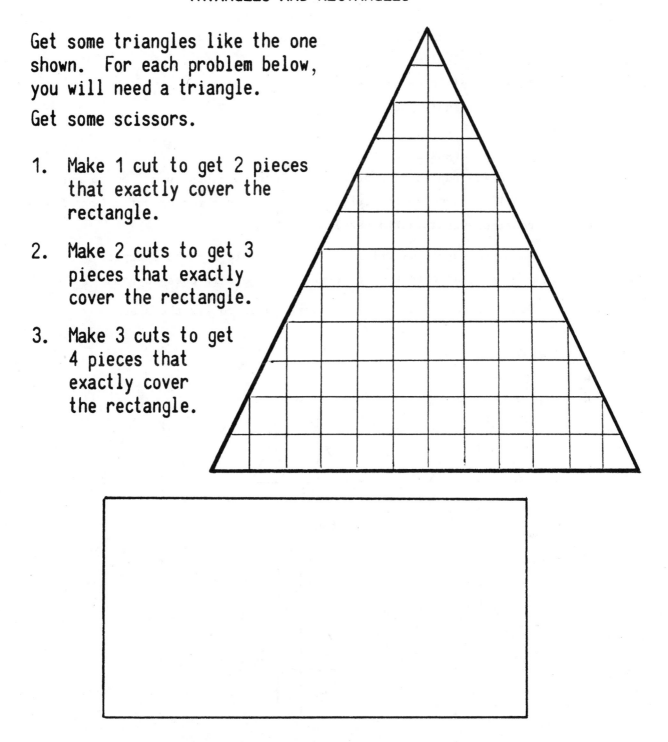

4. What other ways can you find to cut a triangle to exactly cover the rectangle. Record your solutions.

Triangles And Rectangles

Problem-solving skills pupils <u>might</u> use:
- . Use a physical model.
- . Be aware of other solutions.
- . Make reasonable estimates.

Comments and suggestions:

- . Encourage pupils to plan ahead. Some thoughts about how to cut should be done so the cutting is not just a random activity. This requires visualization skills.

- . An extra supply of triangles is advisable for those pupils who make an inadvertent mistake and cut where it does no good.

- . Some pieces may need to be flipped over to fit the space.

Answers:

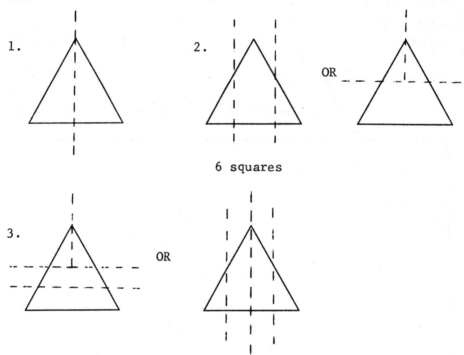

6 squares

4. Many answers are possible.

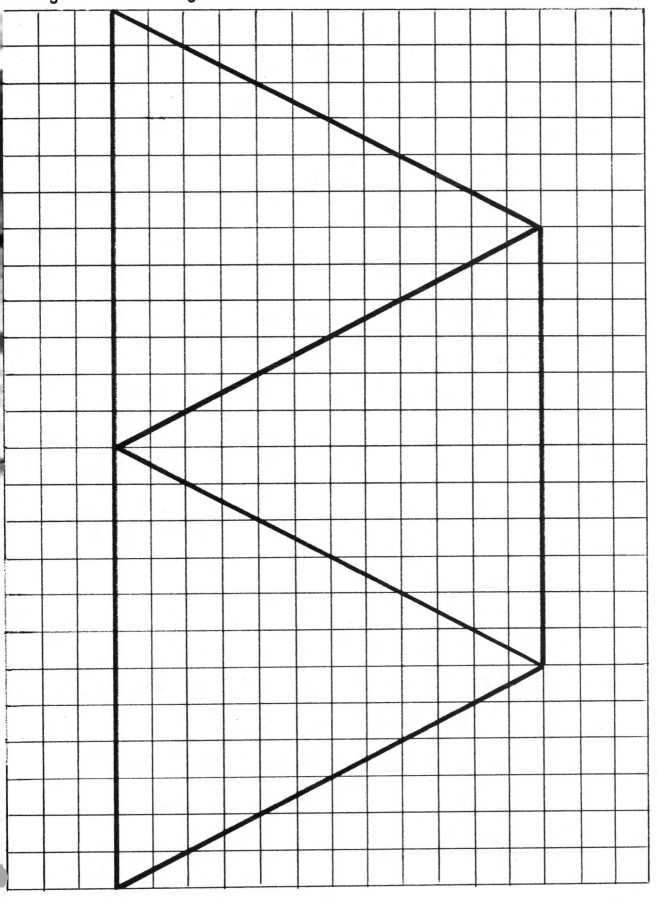

PSM 81

NEIGHBORS

Anne, Betty, and Claudia live next to each other. They work as a chemist, a radio announcer, and a doctor. Find each woman's occupation from these clues.

1. Betty lives in the middle apartment.

2. When Claudia goes away, her cat is fed by the radio announcer.

3. The chemist taps on Anne's wall when her stereo is too loud.

CREATIVE ARTISTS

Alice, Bonnie, Clara, and Donna all practice the creative arts. One of the women is a dancer, one is an artist, one is a singer, and the other is a writer. Find the interest of each woman using these clues.

1. Alice and Clara listened while the singer made her debut.

2. Both Bonnie and the writer have had their portraits done by the artist.

3. The writer, whose biography of Donna is a best-seller, is planning a biography of Alice.

4. Alice and Clara do not know each other.

Neighbors And Creative Artists

Problem-solving skills pupils <u>might</u> use:

. Break a problem into manageable parts.

. Make and use a table.

. Eliminate possibilities using contradictions.

. Guess and check.

Comments and suggestions:

. Logic problems lend themselves to using a table. List one condition such as names across the top and the other condition, occupation, along the side. Use of yes or no or X and O permits the problem solver to use each clue to find the answer.

	Anne	Betty	Claudia
Chemist		Yes	
Announcer		No	No
Doctor		No	

. Some pupils will use guess and check to solve the problems. You might stress that learning to use a table will help on more complicated problems.

Answers:

	Anne	Betty	Claudia
Chemist	No	Yes	No
Announcer	Yes	No	No
Doctor	No	No	Yes

	Alice	Bonnie	Clara	Donna
Dancer	Yes	No	No	No
Artist	No	No	No	Yes
Singer	No	Yes	No	No
Writer	No	No	Yes	No

DIVIDING THINGS UP

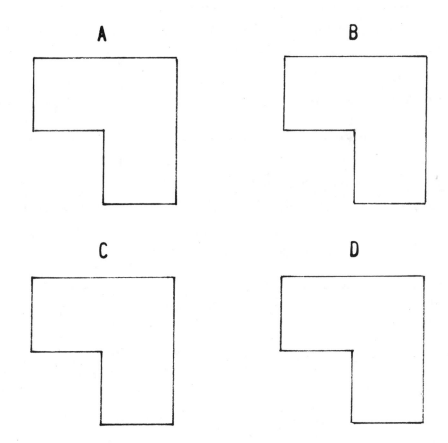

A   B

C   D

1. Divide figure A into two parts exactly the same.

2. Divide figure B into three parts exactly the same.

3. Divide figure C into twelve parts exactly the same.

4. Divide figure D into four parts exactly the same.

5.   Draw two squares so each pig will be in a separate pen.

Dividing Things Up

Problem-solving skills pupils <u>might</u> use:

. Visualize an object from it's drawing.

. Make and use a drawing.

. Guess and check.

. Look at a problem situation from varying points of view.

Comments and suggestions:

. Pupils will likely get solutions for A, B and C quite easily. Part D will be difficult. A possible hint would be to tell pupils that the four parts look like the figure shown, except each is smaller.

. For problem 5 pupils need to "break out" of the idea that squares must be positioned in the normal way.

Answers:

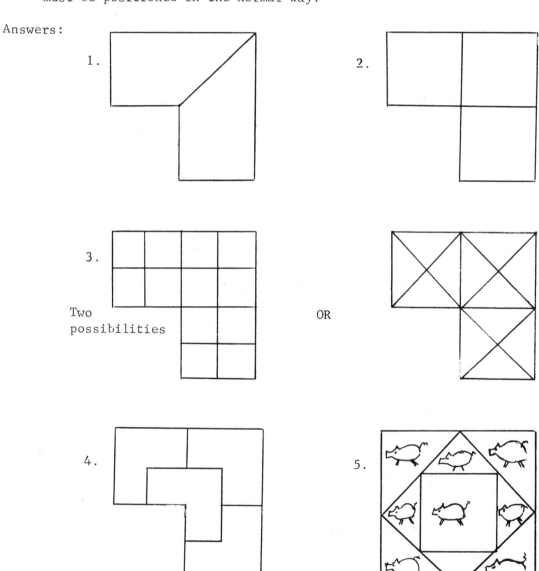

1.

2.

3.

Two possibilities

OR

4.

5.

ADDITION CHALLENGE

1. Use the digits 1 through 9 once each. Fill in the circles
    to make the sum of 999.

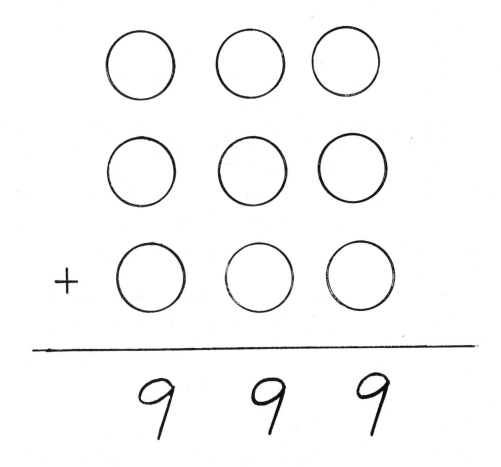

2. Find other solutions. Record.

3. Is it possible to find a solution without using 1 in the
    hundreds column?

4. Use the digits 1 through 9 once each in the circles to make
    the sum 777.

Addition Challenge

Problem-solving skills pupils <u>might</u> use:

- Break a problem into manageable parts.
- Eliminate possibilities.
- Guess and check.

Comments and suggestions:

- See the dialogue in the teacher commentary at the beginning of this section.
- Most pupils will use the guess and check skill. Markers, with the digits 1-9, allow pupils to try possibilities. This procedure avoids a lot of erasing of wrong guesses.
- The digits in the hundreds column control the problem. Only two cases are possible, 1, 3 and 4 or 1, 2 and 5.
- Problem 4 is impossible. Some pupils may spend a lot of time trying to find an answer and may become frustrated. Encourage them to spend a reasonable amount of time looking for a solution and then try to find reasons why a problem can't be answered.

Answers:

1. and 2. Five "basic" solutions are possible.

a.		b.		c.		d.		e.	
	195		159		194		139		169
	476		462		267		286		287
+	328	+	378	+	538	+	574	+	543
	999		999		999		999		999

Many variations of these are possible simply by moving numbers around within a column.

3. No, the only way to get a 9 without using a 1 is to use 2, 3 and 4. But any possibility in the tens column would cause a carry of 1 making the sum 10 instead of 9.

4. Impossible.

PATTERNS IN GEOMETRY

Rectangular Numbers

1st   2nd   3rd   4th

Draw dots to show the next rectangular number.

Place the rectangular numbers in the table below. Then fill in the rest of the table.

1st	
2nd	6
3rd	
4th	
5th	
6th	
7th	
8th	
⋮	
25th	
⋮	
100th	

Triangular Numbers

1st   2nd   3rd   4th

Draw dots to show the next triangular number.

Place the triangular numbers in the table below. Then fill in the rest of the table.

1st	
2nd	
3rd	
4th	10
5th	
6th	
7th	
8th	
⋮	
25th	
⋮	
100th	

Patterns In Geometry

Problem-solving skills pupils <u>might</u> use:

. Make and use a drawing
. Make a systematic list.
. Look for patterns, then make predictions.

Comments and suggestions:

. Some pupils may make each drawing to be certain of their answers.

. The systematic list makes looking for a pattern easier. Some pupils will see the rectangular numbers as a product of two successive numbers, e.g. the fifth rectangular number is 5 x 6 = <u>30</u>. Others will see a pattern in the differences between successive terms as shown below.

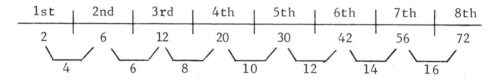

A similar type of pattern can be seen in the triangular numbers.

. The activity is designed to show that each triangular number is half of the appropriate rectangular number. This can be demonstrated by taking two triangular number drawings, turning one over, and moving them together.

. Have pupils fold the page on the two dotted lines. An interesting comparison can be seen between the two tables.

Answers:

Rectangular Numbers

1st	2nd	3rd	4th	5th	6th	7th	8th	...	25th	...	100th
2	6	12	20	30	42	56	72		650		10100

Triangular Numbers

1st	2nd	3rd	4th	5th	6th	7th	8th	...	25th	...	100th
1	3	6	10	15	21	28	36		325		5050

MOVING DIGITS

1. Cut out the digits at the bottom of the page.

2. Place the digits 1, 2, 3, 4, 5, 6 in the squares so no consecutive digits are connected by a line. For example, ③————② is not allowed. Record your solution.

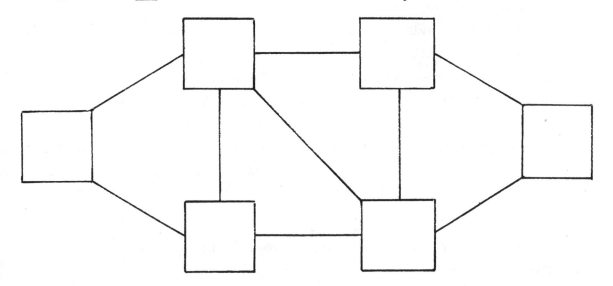

3. Place the digits 1, 2, 3, 4, 5, 6, 7, 8 in the squares so no consecutive digits are connected by a line. Record your solution.

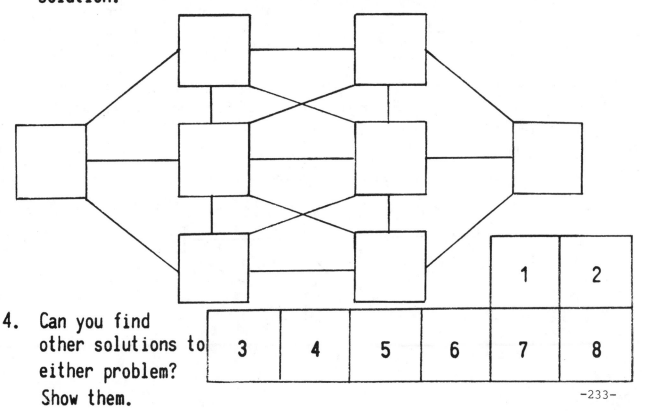

4. Can you find other solutions to either problem? Show them.

				1	2
3	4	5	6	7	8

Moving Digits

Problem-solving skills pupils might use:
. Use a physical model.
. Eliminate possibilities.
. Guess and check.

Comments and suggestions:
. Pupils may need review on the meaning of consecutive digits.
. The use of the digits eliminates the erasure of errors. A correct
solution is recorded only after it has been found.
. In both problems two squares are the keys. Each is connected to
enough other squares that only 1 and 6 (or 1 and 8) can be placed in
those key squares. Placing these two numbers eliminates many possibilities for placing the 2 and 5 (or 2 and 7).

Answers:

2. Two solutions are possible depending on the placement of 1 and 6.
They are rotations of one another.

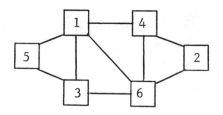 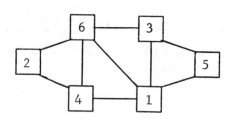

3. Several solutions are possible. The 1 and 8 are placed first, then
the 2 and 7. Next either the 3 and 4 or the 5 and 6 are placed, and
then the remaining two numbers.

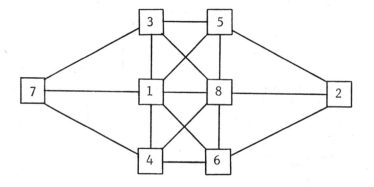

4. See the suggestions for numbers 2 and 3.

IMAGINING FOLDS

Cut out each of these shapes. Imagine that each shape is folded so point 1 lies on point 2.

    a. Draw an accurate sketch of what the folded shape will look like.

    b. Fold the shape. Compare it to your sketch.

1.
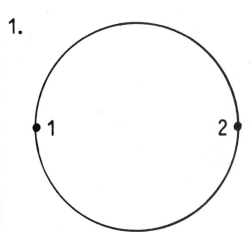

2.

3.
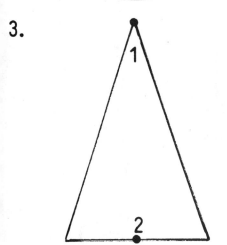

4.
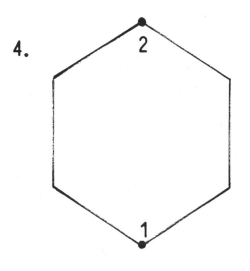

5. Trace and cut out each shape two more times. Keep point 1 in the the same place. Move point 2 as indicated below. Repeat steps

                                                                       a and b.

  a.

                          c.

  b.
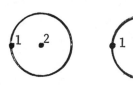

                          d.

-235-

Imagining Folds

Problem-solving skills pupils <u>might</u> use:
- Use a physical model.
- Visualize an object from its description.

Comments and suggestions:
- Visualizing objects is a skill most pupils are weak in. This activity provides an opportunity to predict, then check, what happens when an object is folded. Be sure pupils make a sketch before doing the actual folding.

Answers:

1.   2.   3.   4.

5. a.   b.

   c.   d.

USING CLUES FOR NUMBERS

Use the clues to decide which letter goes with which number.
Write the letter above the number.

1. 10 15 5 M + N = P N > M	**2.** 33 44 55 Z < X Y ÷ 6 > 9
3. 36 4 12 B > A C > 25	**4.** 8 6 4 2 A is not the smallest. R < Y A + M = R
5. 33 20 55 80 P + R < Q P + Q < S	**6.** 103 66 41 13 A < T Y < B A > 41
7. 96 64 50 32 26 A – R > 10 100 – M < 5 T – Y = 6 R < A	**8.** 115 120 125 130 135 B is the largest. C + D = 2 x A E + A = 2 x C

9.   Invent a problem of your own.   Use at least 3 clues.

<u>Using</u> <u>Clues</u> <u>For</u> <u>Numbers</u>

Problem-solving skills pupils <u>might</u> use:
. Guess and check.
. Eliminate possibilities.
. Recognize attributes of a number.

Comments and suggestions:
. Pupils need to know the meaning of the symbols $<$ and $>$. It is helpful to understand that the relationship can be read in both directions. $N > M$ means N is greater than M but it also means that M is less than N. For some problems it may be helpful to rewrite some statements so all the relationships read the same; e.g. in (6), $T > A$, $B > Y$, and $A > 41$.

. Problem 8 has two correct answers. Encourage pupils to find both.

Answers:

1. N=10, P=15, M=5

2. Z=33, X=44, Y=55

3. C=36, A=4, B=12

4. Y=8, R=6, A=4, M=2

5. R=33, P=20, Q=55, S=80

6. T=103, A=66, B=41, Y=13

7. M=96, A=64, R=50, T=32, Y=26

8. D=115, A=120, C=125, E=130, B=135

or

E=115, C=120, A=125, D=130 B=135

CIRCLING NUMBERS

1. Draw circles around any two of these four numbers.

   1   a. Which two did you circle? _____

2   3   b. Write other pairs of numbers you could have
               circled.
   4

   _____ _____ _____ _____ _____

           c. How many different pairs did you find altogether?

                                                    _____

2.   1   How many different pairs of numbers
         2   3   could be circled in this diagram?
       4   5   6
                       Make an organized list showing all
         7   8   the pairs.
              9

3.   1   Without listing, how many different
          2   3   pairs of numbers could be circled
        4   5   6   in this diagram?
      7   8   9   10
        11   12   13   _____
          14   15
             16

<u>Circling</u> <u>Numbers</u>

Problem-solving skills pupils <u>might</u> use:

. Make a systematic list.

. Look for patterns, then make predictions.

. Solve an easier, related problem.

Comments and suggestions:

. Encourage pupils to be systematic about their lists. One scheme is to keep one number of the pair consistent while changing the other number. For example, in problem 2, pupils might start with these pairs, (1,2) (1,3) (1,4) (1,5) (1,6) (1,7) (1,8) (1,9). Then those pairs with 2 could be listed--(2,3) (2,4) (2,5) (2,6) (2,7) (2,8) (2.9).

. To answer problem 3 pupils might say there are 15 pairs with 1 as the first number, 14 pairs with 2 as the first number, ..., 1 pair with 8 as the first number and then add 15 + 14 + ... + 2 + 1 to get the answer.

. Another method for problem 3 involves solving several simpler problems. A table as shown can be developed and used to find the answer.

Numbers	2	3	4	5	6	7	8	9	...	16
Pairs	1	3	6	10	15	21	28	36	...	120

      2   3   4   5   6   7   8   15

Answers:

1. 1,2   2,3   3,4
    1,3   2,4
    1,4

        6 pairs

2. Notice how a vertical listing aids in recognition of patterns.

    1,2   2,3   3,4   4,5   5,6   6,7   7,8   8,9
    1,3   2,4   3,5   4,6   5,7   6,8   7,9
    1,4   2,5   3,6   4,7   5,8   6,9
    1,5   2,6   3,7   4,8   5,9
    1,6   2,7   3,8   4,9
    1,7   2,8   3,9
    1,8   2,9
    1,9   36 pairs

3. 120 pairs

THE CAT AND THE MICE

A cat dreams about being surrounded by 8 mice--7 black ones and 1 white one. After eating one mouse, the cat decided to eat every eighth mouse by going around the circle in the same direction. The white mouse was the <u>last</u> one eaten.

1. Which mouse should the cat eat first? _____

2. If there are 9 mice--8 black and 1 white and the cat eats every ninth mouse, which one should be eaten first? _____

3. How about 12 mice with the cat eating every twelfth mouse? _____

The Cat And The Mice

Problem-solving skills pupils <u>might</u> use:

  . Guess and check.

  . Use a drawing.

  . Use a physical model.

  . Work backwards.

Comments and suggestions:

  . Pupils will probably use guess and check. After the white mouse
    has been eaten, refinement of the guess might be necessary.

  . For consistency with the answers, suggest that pupils label the
    white mouse as mouse #1. Clockwise labeling the mice 1, 2, ..., 8
    or A, B, ..., H will aid in record-keeping. Markers or coins to
    cover a mouse as it is eaten will avoid the situation of having
    to erase incorrect attempts.

  . A group of eight pupils could cooperatively simulate the activity
    by having all pupils standing and then sitting down as "the cat
    ate the appropriate mouse."

Answers:

Designating the white mouse as mouse 1 and moving in a clockwise
direction, the following mouse should be eaten first.

1. 5th mouse

2. 2nd mouse

3. 2nd mouse

There appears to be no pattern one can use to predict the first
mouse to be eaten. Perhaps your pupils can discover a pattern.

THOSE GOOD OLD PRICES

1. Alicia, Bonnie, and Clyde each had a nickel to spend on
    candy. Lollipops are 3 for 1¢; chocolates are 4 for 1¢;
    and jujubes are 5 for 1¢. Each person spent exactly 5¢
    and got 20 pieces of candy. Each person bought some-
    thing different. What did Alicia, Bonnie, and Clyde buy?

2. Slim purchased some of each of these school supplies: paper
    at 2 sheets for 1¢, pens at 1¢ each, pencils at 2 for
    5¢, and erasers at 5¢ each. Slim bought 25 items
    altogether for 25¢. How many of each did he buy?

Those Good Old Prices

Problem-solving skills pupils <u>might</u> use:

. Guess and check.

. Make a systematic list.

. Record solution possibilities.

Comments and suggestions:

. Guess and check may be used by pupils. They should soon discover this is a tedious method for these particular problems because there are many possibilities to try.

. The systematic list works well but pupils may need help in starting the list. In problem 1 a pupil might begin with 5 sets of jujubes. But this makes 25 pieces of candy; 4 sets of jujubes only makes 4¢ spent; 3 sets of jujubes gives 15 pieces of candy but no combination of chocolates and lollipops will bring the total to 20 pieces of candy. Finally 2 sets, 1 set and no jujubes gives the solutions. See the answers below.

. In problem 2, 1 set of each item costs 12¢ and gives 6 individual items. To get the price to 25¢ and the number of items to 25 requires a lot of paper at 2 sheets for 1¢. The answer below shows how the table leads to the correct solution.

Answers:

1.

jujubes	chocolates	lollipops	Total number	cost
5	—	—	25	5¢
4	—	—	20	4¢
3	—	—	won't work	
2	1	2	20	5¢
1	3	1	20	5¢
—	5	—	20	5¢

2.

Erasers	Pencils	Pens	Paper	Total number	Cost
1	2	1	2	6	12¢
1	2	1	28	32	25¢
1	2	2	26	31	25¢
1	2	3	24	30	25¢
⋮	⋮	⋮	⋮	⋮	⋮
1	2	8	14	25	25¢

MR. PAGE'S PUZZLES

1. Mr. Page liked to make up puzzles.

   "If you divide my age by 2," he said, "you'll get a remainder of 1."

   "Furthermore, if you divide my age by 3, 4, and 5, you'll also get a remainder of 1."

   What is the age of Mr. Page?

2. Mr. Page has a certain amount of money in his pocket.

   "If you divide the amount of money I have by 2," he said, "you will get no remainder."

   "Furthermore, if you divide this amount by 3, 4, and 5, you will also get no remainder."

   "But, if you divide the amount by 7, you'll get a remainder of 1."

   What's the smallest amount Mr. Page could have in his pocket?

EXTENSION

Below is an answer to one of Mr. Page's collections of puzzles. The puzzle is like the ones given above. See if you can think of Mr. Page's puzzle.

Answer: 71¢

<u>Mr. Page's Puzzles</u>

Problem-solving skills pupils <u>might</u> use:

. Guess and check.

. Make a systematic list.

. Eliminate possibilities.

. Satisfy one condition at a time.

Comments and suggestions:

. Pupils will look at the first clue, satisfying the condition of
a remainder of 1 when dividing by 2. It may be better, because
fewer numbers work, to first satisfy the condition of a remainder
of 1 when dividing by 5.

Answers:

1. 61 years old.

2. $1.20

Extension: Answers will vary. One puzzle might be:

Find the smallest number that has a remainder of

1 when you divide by 2,

2 when you divide by 3,

3 when you divide by 4,

1 when you divide by 5, and

1 when you divide by 7.

WHAT'S THE COST?

1. How much is the milkshake?
   Use these clues to find out.

   - You can use exactly 7 coins.

   - You can use pennies, nickels, dimes, and quarters.

   - The cost is between $.75 and $1.00.

   Find as many different costs as you can. (There are more than ten.)

2. How much is one cola?
   Use these clues to find out.

   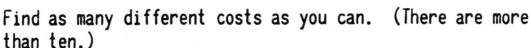

   - The cost is between 20¢ and 30¢

   - You can buy one cola with exactly 6 coins.

   - You can buy two colas with exactly 6 coins.

   - You can buy three colas with exactly 6 coins.

   - You can use pennies, nickels, dimes, or quarters.

What's The Cost?

Problem-solving skills pupils might use:

  . Make a systematic list.

  . Guess and check.

  . Eliminate possibilities.

Comments and suggestions:

  . Be sure pupils understand they have only quarters, dimes, nickels
    and pennies to work with. As an extension, using a half-dollar
    will allow a few more costs to serve as answers.

  . Pupils will probably use one of two strategies. One is to make
    a systematic list showing possible combinations of seven coins.
    The other strategy is to pick a cost and see if that cost can be
    obtained with seven coins; eliminating those that don't work;
    and then trying other costs.

Answers:

1. 17 different costs are possible; 95¢, 91¢, 80¢
    and 76¢ are possible in two different ways.

25¢	10¢	5¢	1¢	Cost
3	2		2	97
3	1	2	1	96
3	1	1	2	92
3	1		3	88
3		4		95
3		3	1	91
3		2	2	87
3		1	3	83
3			4	79
2	4	1		95
2	4		1	91
2	3	2		90
2	3	1	1	86
2	3		2	82
2	2	1	2	77
2	1	4		80
2	1	3	1	76
2		5		75
1	6			85
1	5	1		80
1	5		1	76

2. 24¢   6 coins

24¢

10¢ 10¢ 1¢ 1¢ 1¢ 1¢

48¢

25¢ 10¢ 10¢ 1¢ 1¢ 1¢

72¢

25¢ 25¢ 10¢ 10¢ 1¢ 1¢

-248-

A BARREL OF FUN!

Beatrice Bloom received 6 numbered barrels for her birthday. She thought it would be fun to see how many different ways she could stack them. Show the different ways she can stack them so no barrel has a smaller number directly below it.

EXTENSION

Show how she can stack them so no barrel has a smaller number directly below it or directly to the right of it.

A Barrel Of Fun!

Problem-solving skills pupils **might** use:

. Guess and check.

. Use a physical model.

. Solve an easier problem.

Comments and suggestions:

. Six markers labeled 1, 2, 3, 4, 5, and 6 will be helpful for pupils to try different combinations.

. Many pupils will use a guess and check method. They should notice that barrel 1 must always be at the top and barrel 6 must always be at the bottom.

. The problem can be solved by concentrating on only the first stack. As long as the barrels in the first stack are stacked properly, the remaining barrels can be stacked properly in the second stack. Although very sophisticated, this technique should be pointed out to pupils.

. Let pupils decide whether

```
1  4
2  5    is the same as or different from
3  6
```

```
4  1
5  2  .
6  3
```

Answers:

Notice the systematic way the first stack is found.

```
*1  4        *1  3        *1  3         1  3
 2  5         2  5         2  4         2  4
 3  6         4  6         5  6         6  5


*1  2        *1  2         1  2
 3  5         3  4         3  4
 4  6         5  6         6  5


 1  2         1  2
 4  3         4  3
 5  6         6  5


 1  2
 5  3
 6  4
```

Extension: The answers marked with an asterisk satisfy the extension.

-250-